T0239450

Indian Institute of Metals Series

About the Book Series:

The study of metallurgy and materials science is vital for developing advanced materials for diverse applications. In the last decade, the progress in this field has been rapid and extensive, giving us a new array of materials, with a wide range of applications, and a variety of possibilities for processing and characterizing the materials. In order to make this growing volume of knowledge available, an initiative to publish a series of books in Metallurgy and Materials Science was taken during the Diamond Jubilee year of the Indian Institute of Metals (IIM) in the year 2006. Ten years later the series is now published in partnership with Springer.

This book series publishes different categories of publications: textbooks to satisfy the requirements of students and beginners in the field, monographs on select topics by experts in the field, professional books to cater to the needs of practicing engineers, and proceedings of select international conferences organized by IIM after mandatory peer review. The series publishes across all areas of materials sciences and metallurgy. An eminent panel of international and national experts acts as the advisory body in overseeing the selection of topics, important areas to be covered, and the selection of contributing authors.

Gunturi Venkata Sitarama Sastry

Microstructural Characterisation Techniques

Gunturi Venkata Sitarama Sastry
Formerly, Professor
Department of Metallurgical Engineering
and Dean of Academic Affairs
Indian Institute of Technology BHU
Varanasi, India

ISSN 2509-6400   ISSN 2509-6419 (electronic)
Indian Institute of Metals Series
ISBN 978-981-19-3511-4   ISBN 978-981-19-3509-1 (eBook)
https://doi.org/10.1007/978-981-19-3509-1

This Springer imprint is published by the registered company Springer Nature Singapore Pte Ltd.
The registered company address is: 152 Beach Road, #21-01/04 Gateway East, Singapore 189721,
Singapore

In loving memory of my mother, who patiently waited for my return from Kaasi, but couldn't see my final return. I also dedicate this to my father who constantly encourages me.

Series Editor's Preface

The Indian Institute of Metals Series is an institutional partnership series focusing on metallurgy and materials science and engineering.

About the Indian Institute of Metals

The Indian Institute of Metals (IIM) is a premier professional body (since 1947) representing an eminent and dynamic group of metallurgists and materials scientists and engineers from R&D institutions, academia, and industry, mostly from India. It is a registered professional institute with the primary objective of promoting and advancing the study and practice of the science and technology of metals, alloys, and novel materials. The institute is actively engaged in promoting academia–research and institute–industry interactions.

Genesis and History of the Series

The study of metallurgy and materials science and engineering is vital for developing advanced materials for diverse applications. In the last decade, the progress in this field has been rapid and extensive, giving us a new array of materials, with a wide range of applications and a variety of possibilities for processing and characterizing the materials. In order to make this growing volume of knowledge available, an initiative to publish a series of books in metallurgy and materials science and engineering was taken during the Diamond Jubilee year of the Indian Institute of Metals (IIM) in the year 2006. IIM entered into a partnership with Universities Press, Hyderabad, and, as part of the IIM book series, 11 books were published, and a number of these have been co-published by CRC Press, USA. The books were authored by eminent professionals in academia, industry, and R&D with outstanding background in their respective domains, thus generating unique resources of validated expertise of interest in metallurgy. The international character of the authors and editors has enabled the books to command national and global readership. This book series includes different categories of publications: textbooks to satisfy the requirements of undergraduates and beginners in the field, monographs on selected topics by experts in the field, and proceedings of selected international conferences organized by IIM, after mandatory peer review. An eminent panel of international and national experts constitutes the advisory body in overseeing the selection of topics, important areas to be covered, in the books and the selection of contributing authors.

Current Series Information

To increase the readership and to ensure wide dissemination among global readers, this new chapter of the series has been initiated with Springer in the year 2016. The goal is to continue publishing high-value content on metallurgy and materials science and engineering, focusing on current trends and applications. So far, five important books on state of the art in metallurgy and materials science and engineering have been published and, during this year, two more books are released during IIM-ATM 2022. Readers who are interested in writing books for the Series may contact the Series Editor-in-Chief, Dr. U. Kamachi Mudali, Former President of IIM and Vice Chancellor of VIT Bhopal University at ukmudali1@gmail.com, vc@vitbhopal.ac.in or the Springer Editorial Director, Ms. Swati Meherishi at swati.meherishi@springer.com.

Dr. U. Kamachi Mudali
Editor-in-Chief
Series in Metallurgy and Materials
Engineering

The original version of this book was inadvertently published without the "Dedication" section in the front matter. This section has been included in the book.

Foreword

Materials characterisation plays a central role in the processing and development of materials in general, and more specifically in developing advanced materials. Thus, it becomes necessary to expose students to these different techniques at an early stage of their career. No wonder, this is a required course in most (if not all) metallurgy and materials science programmes both at the undergraduate and graduate levels.

Materials characterisation involves characterising the material for its crystal structure, microstructure and properties. Crystal structure determination is generally covered in a course on X-ray diffraction (although electron microscopy techniques can also be used), while studying the properties of materials is done in more than one course, the mechanical properties form an important component of such a suite. Microstructural characterisation is perhaps the most critical and important of all these, the subject matter of the present book by Professor G. V. S. Sastry.

Microstructural characterisation is typically done using microscopy techniques—optical, scanning electron and transmission electron microscopy or other advanced microscopy techniques to unravel the structural features of materials at different levels of resolution. One can observe the microstructural details at the very gross level using a magnifying glass, or on a finer scale using an optical microscope, and at still finer levels using a scanning electron or transmission electron microscope, or at an atomic scale using a field ion microscope. The undergraduate students are not always exposed to all the advanced techniques, even though they are expected at least to understand their basic principles. The students should, however, be fully familiar with all the basic microstructural characterisation techniques. Professor Sastry provides an outstanding account of the principles, operation and usage of the different types of microscopes offering examples of the type of information that can be obtained along with ways of interpreting the micrographs and diffraction patterns.

There appear to be a few special features of this book. One is a chapter on Fourier Analysis and Fourier Transformations, which is typically not found in undergraduate textbooks (and not even in graduate textbooks). Further, optical microscopy is normally covered first and the concepts of electron microscopes are introduced to overcome the limitations of resolution limit available in an optical microscope. Another special feature of the book appears to be that Professor Sastry decided to provide most of the micrographs and examples based on the work done in India or by persons of Indian origin from overseas.

Professor G. V. S. Sastry is an accomplished teacher with long experience. He has been a practicing electron microscopist for nearly 40 years and the present textbook reflects the experience of the author while explaining some of the important concepts of microscopy.

The University in Varanasi has been a mecca for microscopy for almost 50 years. The first summer school on Metallography was conducted there in 1971, followed by several others on Electron Microscopy and Field Ion Microscopy during subsequent years. Therefore, it is not surprising that the first undergraduate textbook on Microstructural Characterization Techniques has come out from Varanasi.

I am confident that this book will be warmly welcomed by the students and faculty alike. I wish the book a great success.

Orlando, FL, USA   C. Suryanarayana
July 2021

Acknowledgements

My sincere thanks are due to those undergraduate students of IIT(BHU), Varanasi, formerly IT-BHU, Varanasi, whose probing questions and queries helped me a great deal in improving the delivery of my Lectures for a Course on Metallographic Techniques. I am thankful to all those post-graduate students who chose Materials Characterisation Techniques, and the post-graduate students and doctoral students who opted for Advanced X-ray and Electron Microscopy for their unquenchable thirst for deeper knowledge about the interference of electron waves and about what happens to their charge when they do so etc. They were intrigued by the fact that electrons do not repel each other when they are focussed to a spot. Such in-depth discussions happened to be an indirect motivation for me to write this book. A key role is played by Prof. C. Suryanarayana, mentor for my doctoral study, in all the knowledge and experience I gained in electron microscopy. He used to answer untiringly all my endless queries about electron microscopy and electron diffraction. Our long discussions helped me pick up the threads in the initial years. I owe a great deal to him for making me what I am in the field of electron microscopy.

The textbooks available for students in the early eighties were the classics, Electron Microscopy of Thin Crystals by Hirsch et al. (Electron microscopy of thin crystals by P. B. Hirsch, A. Howie, R. B. Nicholson, D. W. Pashley and M. J. Whelan, Butterworth, London,1965) and Fundamentals of Transmission Electron Microscopy by Heidenreich (Fundamentals of Transmission Electron Microscopy, R. D. Heidenreich, Interscience Publishers Inc.; First Edition (January 1, 1964)). The rigor of treatment in these two books was beyond the comprehension of undergraduates, I felt. However, I couldn't sit down to write the present book at that time obviously.

During my doctoral research, I was fortunate to have interaction with Prof. G. Van Tendeloo, Reijks Universitaire Centrum, Antwerpen, Belgium, who was on a visiting assignment to our department. He introduced me to the techniques of high-resolution electron microscopy and structure imaging. I am indeed grateful to him for the same. I must acknowledge the companionship and competition provided in my initial years at the microscope by my fellow doctoral researchers, Dr. Fawzy Hosney Samuel and Dr. Jogender Singh. We used to compete for time on Philips EM300, and in the process, I learnt the skills of interpreting the diffraction patterns on the screen of the microscope itself so that I could get maximum results out of one sitting at the microscope. In retrospect, this competition helped me and I am thankful to them for this. I realised soon that it is also important to know about the functioning of the microscope, its ancillaries and its alignment. I gained this knowledge from Sri S. Krishnamurty and Sri S. N. Lal, the Scientific Officers, who helped me troubleshoot whenever the microscopes were either down or malfunctioning. It was a joy working with them. In recent years, it is Sri Lalit Kumar Singh who benefited from this experience and also Sri Ashok. I am greatly indebted to Sri Lalit Kumar Singh who always readily helped me by mailing many micrographs from my archives at the Department, during the last 1 year.

The Department of Metallurgical Engineering, IIT(BHU), Varanasi, has stalwarts in the area of Materials Characterisation like Professors T. R. Anantharaman, P. Ramarao and S. Ranganathan, P. Ramachandra Rao, S. Lele, D. S. Sarma, with whom direct or indirect interactions helped me in many ways. I sincerely and reverentially acknowledge the same. Perhaps the closest interaction that I am fortunate to have in my age group is with Prof. K. Chattopadhyay and I continue to benefit from it. I sincerely acknowledge the inspiration I derive from him. Acknowledged electron microscopists of the country such as Dr. Srikumar Banerjee, Prof. O. N. Srivastava, Prof. Dipankar Banerjee, Dr. V. S. Raghunathan and Dr. M. Vijayalakshmi with whom I had closer interactions have also inspired me. I am grateful to Dr. Peter Wilbrandt, Institüt fur Metallforschung, Universität Göttingen, Germany, for acquainting me with some of the nuances of high-resolution electron microscopy.

I am very much thankful to the Publications Committee of the Indian Institute of Metals, Prof. B. S. Murty, Prof. K. Bhanushankara Rao and Dr. U. Kamachi Mudali, in particular, to have readily accepted to publish my book, and also Ms. Swati Meherishi from Springer to patiently wait for me to file-in the chapters. Special thanks are due to Dr. N. Eswara Prasad for his valuable support. The IIT(BHU), Varanasi, provided me initial support through an offer of Institute Professorship, when I began writing this book. I gratefully acknowledge the same. Professor Srikant Lele and Professor C.Suryanarayana were kind enough to give their valuable suggestions. The former read through the initial chapters of the manuscript, while the latter went through the completed book. Professor B.N.Sarma helped me in compiling some of the Fourier Transforms using Mathematica.

While writing a book, we often acknowledge all the help and encouragement received from the professionals in the field at the outset. But we know that from day one, it is the wife and the family who take the brunt of it and without whose encouragement and patient waiting, a book could never be completed. I am indeed indebted to my wife Sirisha who also took time off to readout the manuscript for me while I was compiling in Springer Monograph template using Overleaf. My sons and their families also helped me a great deal in retrieval of some files when they went corrupt or resolve some issues with Graphics packages on Linux platform.

I sincerely and reverentially acknowledge the blessings from my spiritual gurus, Swami Ramananda Nadha, Varanasi, and Sahajayogini Mata Sudha Patel, London.

Kakinada, India   Gunturi Venkata Sitarama Sastry

Contents

About the Author

Prof. Gunturi Venkata Sitarama Sastry is former Professor, Department of Metallurgical Engineering and also Dean of Academic Affairs, IIT(BHU), Varanasi, India for a three year period. He began extensive use of transmission electron microscopy since his doctoral work on rapidly quenched aluminum alloys at IT-BHU, Varanasi. His research tools have been diverse microstructural characterization techniques with major emphasis on TEM. Most of his publications in high impact international Journals reflect this vast expertise in the field. Several of his students benefited from his courses on metallographic techniques taught at the Department of Metallurgical Engineering, IIT(BHU), Varanasi, India (formerly Institute of Technology, BHU). He also conducted many short-term courses on electron microscopy and associated techniques at various institutions and research laboratories. He headed the National Electron Microscopy Facility established at the Department for over five years and later worked for the augmentation of new facilities over there. The Electron Microscope Society of India recognized his contributions to the field by bestowing upon him the 'Lifetime Achievement Award 2017'.

Chapter 1
Introduction

1.1 Requirement to 'See'

A large portion of materials characterisation involves microstructural observation and studies. Strangely though, much of the information about the morphology or material behaviour can be understood by characterisation techniques other than visual observation of the microstructure. Perhaps this tendency is related to the innate desire of human mind to always visualise or picturise a concept. Nevertheless, when we consider heterogeneous materials, certain properties such as stiffness, electrical conductivity, etc., not only depend on the relative volume fractions of the phases but also on the connectedness of the constituents of the heterogeneous material. This can also be modelled mathematically but visual observation definitely makes it 'easy' to 'see' the connectedness. We shall elaborate further on this aspect by taking an example of the strengthening behaviour of a steel processed by Severe Plastic Deformation (SPD method). The principle of this technique is to induce very large plastic strain in a metal without affecting the shape and size of the sample. The process is repeated many times till the sample is no longer in a position to accommodate additional induced strain. As a result, a high density of dislocations is created in the sample. Eventually, when the density increases beyond a critical value dynamic recovery takes place leading to the formation of subgrains bound by low-angle boundaries within the deformed grains. When the deformation process is repeated further, dynamic recrystallisation takes place giving rise to ultra-fine-grained structure.

Samples at different stages of deformation are characterised by X-ray diffraction and microhardness measurements. Systematic measurements of microhardness are depicted in Fig. 1.1 as a function of depth from the top surface of the billet and equivalent strain-induced. The hardness plot clearly differentiates the stages of initial build-up of high density of dislocations followed by dynamic recovery and recrystallisation. However, the estimated density of dislocations at each stage of accumulated equivalent strain, their configurations, any possible subcell formation,

Fig. 1.1 A plot of Vicker's hardness against equivalent strain [after Deepa Verma, N. K. Mukhopadhyay, G. V. S. Sastry, and R. Manna, Met. and Mater Trans. 47A (2016) 1803. With Permission]

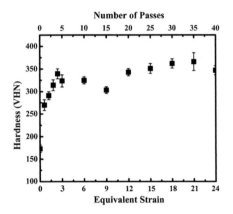

Fig. 1.2 Microstructure of IF steel ECAPed for accumulated strain of 1.8 [after Deepa Verma, N. K. Mukhopadhyay, G. V. S. Sastry, and R. Manna, Met. and Mater Trans. 47A (2016) 1803. With Permission]

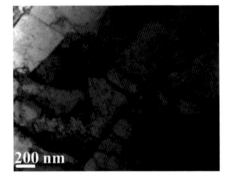

etc., need to be assessed by X-ray line profile analysis. It can provide an accurate estimate of the dislocation density, stored energy and coherently scattering domain size (subcell size) provided the chosen line profile fitting methods can separate the contributions of the particle (domain) size and stored energy to the peak width or Full Width at Half Maximum (FWHM) effectively. Yet another technique, which is based on electron diffraction, called Electron Backscattered Diffraction (EBSD) goes close to the above requirement but only in microregions. This technique brings in an additional benefit of providing texture information. While the information gathered from above-mentioned methods is complimentary to the visual observation of the microstructure, these techniques cannot give location-specific information, e.g. the distribution of the dislocations within a grain or near the grain boundary, (see Fig. 1.2) their mutual interactions and in specific cases their type viz., edge, screw or mixed type.

For gaining such information about (i) dislocations or shapes of grains, (ii) dislocation-particle interactions in case of multi-phase materials and (iii) phase constitution and phase identification one needs a microscope with appropriate resolving power.

1.2 The New Definition

In the modern context, a microscope can be described schematically as shown in the diagram, Fig. 1.3.

The surface or through-thickness microstructure of a chosen sample can be explored using a probe, that was and is ordinary light from the last three centuries, can now be any radiation or tool. Interaction of the probe with the sample can give rise to a signal or response by the sample which may either be transmitted or reflected. The microscope processes this signal and renders an image on a screen (or eye) or creates a visual display on a suitable CRT. When the microscope uses light or electrons as a probe and forms an image obeying the laws of optics, it can be truly

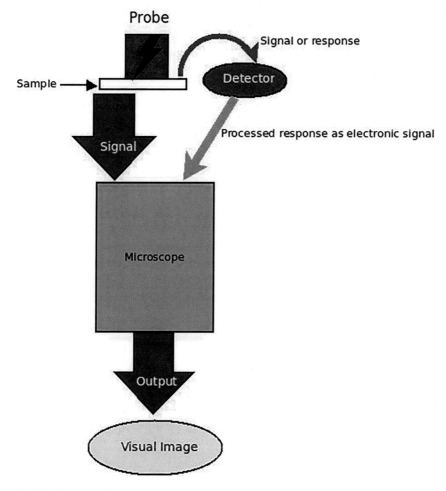

Fig. 1.3 A conceptual modern microscope

called a microscope. Other instruments borrow the generic name of a microscope but do not follow the laws of optics in forming a visual display and hence are not true microscopes. The ones which obey the laws of optics depend on the diffraction phenomenon for rendering the image and hence are designated as 'diffraction limited'. Of course, both systems introduce distortions while processing the signal from the sample. This aspect can be further understood by referring to a systems approach and signal processing theories.

1.3 Scope of the Book

Books often become voluminous when we tend to explain all the aspects of any particular topic. The undergraduate student or even the postgraduate one in her/his first exposure to the field finds herself/himself lost in the detail. This was realised while teaching the undergraduate Course on 'Metallographic Techniques' at the department. The present book, therefore, is designed to fill the gap between the classical books which are very brief while maintaining the rigour and detailed books which almost border on the category of Reference Books. It is composed to take care of the syllabus of the course to the extent possible. The said course covers X-ray diffraction techniques also, which require to be omitted here so as to conform to the title of the book. Readers are suggested to refer to standard books on X-ray crystallography and X-ray diffraction. Certain important topics which are outside the syllabus of the course are also included, e.g. Fourier transformation is an important topic to understand not only the successive formation of diffraction pattern and image in a microscope but also to appreciate the full details observed in a diffraction pattern. Similarly, sufficient emphasis is laid on electron diffraction since a detailed working knowledge of it is key to understand the microstructural features. Though the knowledge of dynamical theory of electron diffraction and resulting image contrast is essential for the correct and complete interpretation of the image, it could not be included as it goes beyond the scope of the book. Contents covered in the present form of the book are sufficient for both UG and PG students of materials engineering and materials science. Examples of diffraction patterns and microstructures are sighted from the work done in India or done by Indian authors elsewhere only to bring to the knowledge of students about the expertise available around, while fully acknowledging the experts in specific areas outside India. Credit to these authors is given by directing the reader to their books and publications for advanced reading. In order to meet the time schedules, only a few exercises that judge the grasp of important concepts are provided at the end of each chapter at present.

Chapter 2
Electromagnetic Waves and Electron Waves

Diffraction plays a pivotal role in electron microscopy as we will see in later chapters. It is a manifestation of their wave nature in addition to their particle nature. But so is the phenomenon called interference. Which of the two is more fundamental has been a big debate for some time. Hence we will briefly discuss this point at first. In later part of the chapter we will discuss the superposition of waves, charged particle optics (electrons in our particular case) and the lenses that are required to refract the electron waves to form either a diffraction pattern or an image.

2.1 Interference and Diffraction

The first question that arises in our mind in understanding the difference is, 'Is there a difference between the two?' This debate is four centuries old and seems to have got resolved in the nineteenth and twentieth century. On a rudimentary level interference can be visualised as superposition of waves while diffraction is the bending of rays around an obstruction.

Diffraction was first observed by Grimaldi in Italy in the seventeenth century. Newton studied and tried to interpret it unsuccessfully on the basis of his particle theory of light. Wave theory was introduced by Fresnel, based on Huygen's principle of secondary wavelets, which was later perfected by Sommerfeld in 1896. Interpretation of light as a set of waves propagating from a point source in all directions with a constant velocity helped in assigning a phase value and an amplitude to them. When a point source is sufficiently far from an obstacle such as a straight edge or a screen with an orifice, the wavefront (envelope of all the points on the waves which are having the same phase) can be assumed to be planar, as shown in Fig. 2.1.

© The Author(s), under exclusive license to Springer Nature Singapore Pte Ltd. 2022   5
G. V. S. Sastry, *Microstructural Characterisation Techniques*, Indian Institute
of Metals Series, https://doi.org/10.1007/978-981-19-3509-1_2

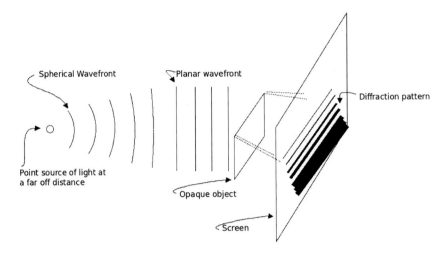

Spherical Wavefront

Planar wavefront

Diffraction pattern

Point source of light at
a far off distance

Opaque object

Screen

Fig. 2.1 Diffraction at straight edge

It is observed that when such planar wavefront meets an obstacle say, a straight
edge, the wavefront instead of moving forward bends over the straight edge. If now
a screen is placed ahead of the straight edge, we observe that the shadow cast on the
screen is a group of straight fringes. The first fringe is also not straight ahead of the
straight edge but slightly lower in position (see Fig. 2.1) indicating that the wavefront
has bent over the edge while crossing the obstacle. The fringes constitute a diffraction
pattern formed by the straight edge and the phenomenon is called diffraction of the
wave.

A rigorous explanation and theoretical formulation of the phenomenon requires
that the planar wave front breaks down into spherical secondary waves as soon as it
touches the straight edge, a concept due to Huygens. Superposition or more precisely
interference, of these secondary waves, leads to the formation of a series of alternate
dark and bright fringes or bands as a shadow; called the diffraction pattern of the
straight edge. Thus the shadow is not a uniformly dark region as intuitively expected,
but a display of the diffraction pattern of the straight edge.

The above observations further stress the point that interference of waves is more
fundamental requisite of a diffraction pattern. A diffracted beam results when there
is constructive interference of the waves in that direction. Interference can take place
in a variety of situations and that leading to diffraction happens to be a special case.

If two waves having the same frequency are travelling in two directions represented by two vectors k_1 and k_2 they may interfere in different ways. When waves are
of light, it is necessary that the two waves are having their electrical vectors oriented
in the same direction, say the plane of the paper, as shown in Fig. 2.2a–c:

(a) The resultant wave will have an amplitude which is twice that of the single wave
    as both waves are in phase and the amplitudes add up. See Fig. 2.2. You may
    notice that the resultant wave is purple, while the two individual waves having

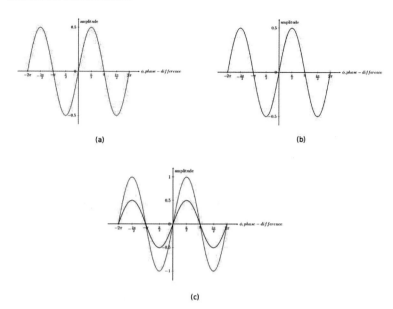

Fig. 2.2   Interference of two waves with the same amplitude and phase

colours red and blue give rise to magenta when superimposed. The colours are chosen in such a way that they give rise to the natural mixed colour to illustrate interference.

(b) In case the amplitude of one wave is less than the other, the resultant wave will have an amplitude which is the algebraic sum of the individual amplitudes of the two waves. Figure 2.3 illustrates such a case where the two waves are also having phase differences in addition to the difference in amplitude.

(c) When the two waves travel with a phase difference, amplitude gets attenuated. If the phase difference is 180°, then complete attenuation takes place whether the waves are travelling in the same or opposite directions. Note that the illustration in Fig. 2.4 shows some residual amplitude of the resultant wave (green colour). This is due to a very small difference in points of intersection of the two waves w.r.t. the x-axis.

You are cautioned not to carry the impression that the above superposition of the 1-D waves with 180° phase difference gives rise to complete cancellation of the waves resulting in zero energy at that instant. In reality, the photons cannot annihilate each other. We can understand this better when we consider that (i) the total energy of the photon is a sum of its kinetic energy and potential energy and that (ii) an optical wave has both electrical and magnetic field vectors associated with its propagation. We will not elaborate on this further.

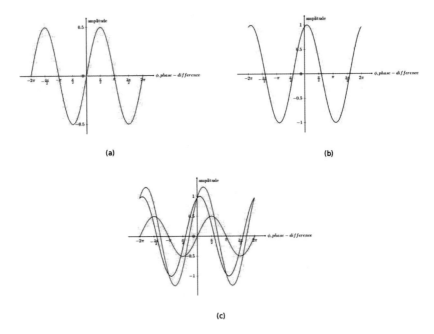

Fig. 2.3 Interference of two waves with different amplitudes and phases

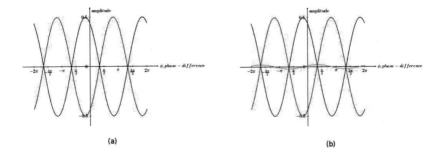

Fig. 2.4 Waves with same amplitude but 180° phase difference

2.2   Charged Particle-Wave Optics

While most of the concepts related to waves discussed in the above section are applicable to charged particle-wave optics, be the particles electrons or ions, there are distinct attributes of charged particles which need to be considered when we want to extend the formulations of geometrical optics of photons to electrons. Since our goal is to understand the optics of transmission electron microscope, we do not discuss the optics of the ion beams separately here. Electrons possess an intrinsic charge, restmass and spin. Therefore, a rigorous treatment of the related optics requires to be done using quantum mechanical wave mechanics. These concepts are worked out in detail by several researchers earlier. For a detailed understanding of the subject and its implications on the design of the microscope the reader is referred to (Rose 2009; Groves 2014; Hawkes and Kasper (1994)).

Electrons possess spin and hence satisfy Dirac's equation (Jagannathan et al. 1989) rather than Schrödinger's equation. In electron optics, however, spin effects are usually negligible except in case of very low voltage scanning electron microscopy (ESEM, Low voltage microscopy of oxide materials). In order to construct the geometrical optics we should be able to compute the precise ray path taken by electrons from their point of generation to their point of focus on the optic axis, by the action of a lens. Information regarding the path taken by an electron is represented mathematically by the integral of a Lagrangian function. The integral, also known as Hamilton's principle function, has an extremum for the actual path taken amongst the many possible paths which are quantum mechanically possible. By this method we get the spatial coordinates of the electrons of a ray path between two points, disregarding its actual time of arrival at any point of time.

Lens can be either an electrostatic field or a magnetic field in the case of electrons since both can refract electrons. The electrostatic lens depending on the type of charge, i.e. positive or negative, can either diverge or converge the electrons passing through them respectively. Magnetic field or a lens with magnetic field lines passing from north pole to south pole of the magnet can only converge the electron beam as the two poles always exist together.

2.3 Electrostatic and Electromagnetic Lenses

2.3.1 Electrostatic Lenses

Uniform, symmetric electric fields can be easily generated between two electrodes that are oppositely charged, cathode and anode or within the central hole of a circular plate that is negatively charged. Electrons that are travelling along the optic axis (OE in Fig. 2.5) experience a radial electrostatic force that is the same in all directions and thus continue along the optic axis without any change in their path. Any other ray (FG in Fig. 2.5) which is travelling at an angle to the optic axis would be repelled towards the optic axis and converge at a point as shown in Fig. 2.5. The electrode when positively charged would attract the electrons leading to a divergent beam of electrons. In this way, it is possible to have both converging and diverging (concave lens in case of light optics) lenses with electrostatic fields. These lenses, nevertheless, are cumbersome and were used in very early generation electron microscopes.

Fig. 2.5 Eletrostatic lens
action

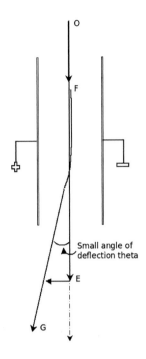

2.3.2   *Electromagnetic Lenses*

A current carrying coil produces a magnetic field at the centre of the coil. If the current is dc, which is the case in practice, the magnetic field is constant. An electron beam passing through the centre of the coil is deflected but not focussed due to the uniform field. In order to get focussing action a short coil is necessary such as that shown in the Fig. 2.6. The axis-symmetric field created in this case is confined to a narrow region. The force exerted by the magnetic field on an electron travelling through this, **F** is given by the simple equation

$$\mathbf{F} = -e(\mathbf{v} \times \mathbf{B}),$$

where e is the electron charge, **v** is the velocity of the electron at any position and **B** is the magnetic induction or field strength.

As the field lines are curved, the velocity **v**, magnetic induction **B** and force exerted by the field **F** are all variables according to position and are thus vector quantities. Since the right-hand side of the equation is a vector product the force exerted would be normal to the plane constituted by **v** and **B**, the angle between these two being any instantaneous angle as **B** varies continuously. Therefore, **F** has no component along the direction of **v** and hence the electron speed remains constant.

Any electron entering the concentrated magnetic field of the lens at an arbitrary angle ϕ from the optic axis experiences force **F** as shown in the Fig. 2.6 and is equal

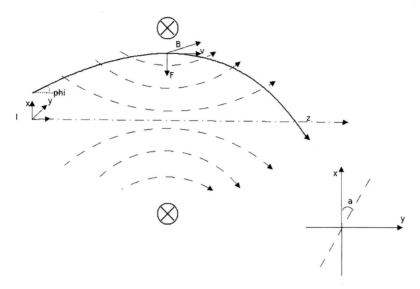

Fig. 2.6 Short coil lens

Fig. 2.7 a Symmetric
magnetic lens, **b** Magnetic
lens with shaped pole pieces

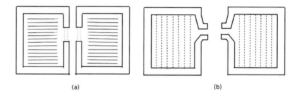

to e **v B** Sin(θ) . As the magnetic induction **B** and **v** continuously change along the
trajectory, so does the force **F** on the electron. Therefore, its path is a helix *i.e.* the
electrons which form the image I on the optic axis z would be rotated by an angle α
with respect to x-coordinate. The electrons that travel parallel to Iz continue to do so
and experience no force since $\theta = 0$ and **F** = 0. When the solenoid is having extended
volume (i.e. cylindrical in shape) it can only exert a focusing action on electrons that
are emanating from a point on the axis of the cylinder, but not on a beam of electrons
passing parallel to the axis of the cylinder. Therefore, the magnetic field line should
be confined to a very narrow area in order to achieve lensing action. This can be
accomplished by encasing the solenoid inside an annular soft iron box. The box has
a small gap in the bore of the annulus as shown in Fig. 2.7a.

A more efficient design adopted in the modern microscopes is to provide ferromagnetic pole pieces shaped and manufactured to a high degree of accuracy, as
shown in Fig. 2.7b. To achieve a strong magnetic field that is capable of focusing
fast electrons, the number of turns in the solenoid is increased by choosing very fine
diameter copper wire. The fine diameter of the wire leads to the generation of large
amount of heat due to the resistive heating. Hence water cooling is also provided

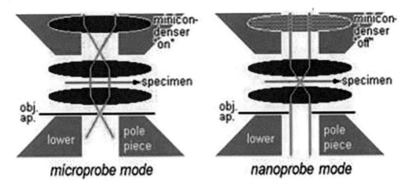

Fig. 2.8 Schematic sectional diagram of a current design incorporating a mini condenser lens (objective pre-field) in the objective lens in Technai 20G^2 TEM [Courtesy of M/s ICON Analytical Pvt. Ltd., Mumbai, India]

around the annular box. The cross-sectional view of an objective lens in a typical modern microscope is given in Fig. 2.8. Depending upon the configuration of the microscope, the design details may vary.

2.3.3 The Lens Parameters

A detailed analysis of the focussing action of a thin magnetic lens is given by Reimer and Kohl (2008), Hawkes and Kasper (2017). Using the approach of these authors, one can derive an expression for the focussing power of a thin magnetic lens as

$$\frac{1}{f} = \left[\frac{e^2}{8m E_o} \right] \int B_z^2 dz \tag{2.1}$$

where f is the focal length of the lens, e is the electron charge, m is the restmass of the electron and B_z is the z-component of the magnetic field. The integral can be simplified for a symmetrical bell-shaped field to $\frac{\pi}{16} a B_o^2$, where 2a is fullwidth at half maximum and B_o its maximum magnetic field at the centre of the lens.

By substituting the terms in the equation one can calculate the f of a magnetic lens for 200 kV accelerating voltage to be 22 mm. Therefore, when the lens is strongly energised, it will have a short focal length and the achieved magnification is small. Compared to this, when the lens is weakly or moderately energised it will give rise to higher magnification. The electromagnetic lenses have this advantage of using the same lens for achieving different magnifications (see Fig. 2.6). Also notice that the angle of convergence may vary and become nearly zero.

Exercises

Q1. Attempt to extend the concepts of constructive and destructive interference to the case of electrons and understand electron diffraction. Hint: Electrons obey quantum mechanics rather than classical wave mechanics.

Q2. Explain the difference between Fraunhofer diffraction and Fresnel diffraction. Relate this to the distance between the diffracting object and the place of observation of the diffraction pattern.

Chapter 3
Fourier Analysis and Fourier Transformation

A superposition of cosine and sine functions with chosen frequencies and phase relationships can represent any periodic or non-periodic function conforming to the shape of a regular body. This was proposed by Fourier, though in some form it was already being used. The study of decomposition of such functions is termed as Fourier analysis. It was a bold idea that led to solutions of other problems in engineering and physics as we often deal with differential equations that are homogeneous. Homogeneous equations or linear equations are those for which a combination of two solutions of the equation is also a solution. There are two types of Fourier expansions, Fourier series and Fourier transforms. We will demonstrate complete methodology and work out examples of some standard shapes, bringing out their relevance to electron diffraction and imaging.

3.1 Fourier Analysis

Fourier theorem states that any periodic function $f(x)$ (it can be a non-periodic function also) can be expressed as sum of a series of Cosine and Sine waves of wavelengths which are integral sub-multiples of the wavelength λ, of $f(x)$. That is

$$f(x) = \frac{1}{2}C_o + C_1 Cos(\frac{2\pi x}{\lambda} + \alpha_1) + C_2 Cos(\frac{2\pi x}{\frac{\lambda}{2}} + \alpha_2) + C_n Cos(\frac{2\pi x}{\frac{\lambda}{n}} + \alpha_n) + \cdots$$

$$(3.1)$$

The first term can also be taken into the series as $\lambda = 0$ and thus n represents the sub-multiples or order or harmonics. C_n associated with each term is the amplitude and a_n a phase angle. The term a_n signifies phase shifts in the chosen trigonometric functions

© The Author(s), under exclusive license to Springer Nature Singapore Pte Ltd. 2022
G. V. S. Sastry, *Microstructural Characterisation Techniques*, Indian Institute
of Metals Series, https://doi.org/10.1007/978-981-19-3509-1_3

that are necessary to match the function f(x). Determination of these quantities for each term of the series is called *Fourier analysis*.

3.1.1   Fourier Coefficients

For calculating the coefficients one can also write a simpler series in Cosine and Sine functions to represent the chosen f(x). Then Eq. (3.1) takes the following form:

$$f(x) = a_o + \sum_n \left[a_n \text{Cos}\left(\frac{2\pi x}{\lambda}\right) + b_n \text{Sin}\left(\frac{2\pi x}{\lambda}\right) \right] \tag{3.2}$$

Once again we invoke certain properties of the trigonometric functions chosen in order to evaluate the coefficients a_o, a_n and b_n. The arguments of the two trigonometric functions have 2π included in them and thus n = 1 terms have wavelength same as f(x), see Fig. 3.1, while n = 2 terms represent waves which are $\frac{1}{2}$ periodic.

Therefore, the RHS function is almost of period matching with that of the original function f(x). None of the terms can overshoot λ. The period can be shorter than λ, say $\frac{\lambda}{2}$, in which case only the even n's have non-zero coefficients. The two trigonometric functions in Eq. (3.2) are such that when we carry out the summation over integral number of wavelengths the net summation is either zero or $\frac{\lambda}{2} \delta_n$ or $\frac{\lambda}{2}\delta_m$ as demonstrated below. Here $\delta_{n,m}$ are Kronecker deltas. Such functions are known as orthogonal functions. Let us take the product of Sine and Cosine functions integrated over an interval with chosen values of n and m.

$$\int_0^\lambda \text{Sin}\left(\frac{2\pi n x}{\lambda}\right) + \text{Cos}\left(\frac{2\pi m x}{\lambda}\right) dx ....(i)$$

This integral can also be expressed as sum of two Sine functions using trigonometric relations as follows:

Fig. 3.1 A representation of fourier series

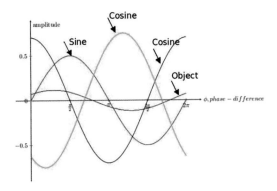

$$\frac{1}{2}\int_0^\lambda \left[Sin(n+m)\frac{2\pi x}{\lambda} + Sin(n-m)\frac{2\pi x}{\lambda} \right] dx....(ii)$$

But this integrates to zero over the interval 0 to λ as both terms undergo integral number of oscillations within this interval. The special case of n = m is also taken care of.

Similarly, when you consider a Cosine product such as

$$\int_0^\lambda Cos(\frac{2\pi nx}{\lambda})Cos(\frac{2\pi mx}{\lambda})dx = \frac{1}{2}\int_0^\lambda \left[Cos(n+m)\frac{2\pi x}{\lambda} + Cos(n-m)\frac{2\pi x}{\lambda} \right]dx ....(iii)$$

This integral also goes to zero over the chosen interval except at n = m. The second term goes to 1 and the integral would be equal to $\frac{\lambda}{2}$. The third product

$$\int_0^\lambda \int_0^\lambda Sin(\frac{2\pi nx}{\lambda})Sin(\frac{2\pi mx}{\lambda})dx = \frac{1}{2}\int_0^\lambda \left[-Cos(n+m)\frac{2\pi x}{\lambda} + Cos(n-m)\frac{2\pi x}{\lambda} \right]dx ....(iv)$$

also integrates to zero except at n = m where again it equals $\frac{\lambda}{2}x\delta_{n,m}$.

Using the orthogonality condition we can arrive at the coefficients a_o, a_n and b_n.
$\int_0^\lambda f(x)dx = \int_0^\lambda a_o dx = a_o x|_0^\lambda = a_o\lambda$ (area under the curve)

$$a_o = \frac{1}{\lambda}\int_o^\lambda f(x)dx \qquad (3.3)$$

$$\int_0^\lambda f(x)Cos(\frac{2\pi mx}{\lambda})dx = a_m\frac{\lambda}{2}$$

$$a_n = \frac{2}{\lambda}\int_0^\lambda f(x)Cos(\frac{2\pi nx}{\lambda})dx \qquad (3.4)$$

$$\int_0^\lambda f(x)Sin(\frac{2\pi mx}{\lambda})dx = b_m\frac{\lambda}{2}$$

$$b_n = \frac{2}{\lambda}\int_0^\lambda f(x)Sin(\frac{2\pi nx}{\lambda})dx \qquad (3.5)$$

In order to improve the fit, we can introduce phase shifts α_n as in Eq. (3.1).

Even and Odd functions:
We have introduced Cosine and Sine functions for writing down the Fourier series, while realising that even functions can only be fitted by Cosines while odd functions can be fitted by Sines. A function is understood to be even if $f(\theta) = f(-\theta)$ and odd otherwise (i.e. $f(\theta) = -f(\theta)$). Therefore certain parity relations can be written down for a_ns and b_ns.

Even function: $a_n = a_{-n}$

Even function, also real: $a_n = (a_{-n})^* = a_{-n}$
here a_n is also real

Odd functions: $a_n = -a_{-n}$

Odd functions, also real: $a_n = (a_{-n})^* = -a_{-n}$
here a_n is purely imaginary

The above follow from the conditions that B_ns $= 0$ for even functions and A_ns $= 0$ for odd functions.

3.1.2 Exponential Series

The trigonometric functions can also be expressed as exponential series and simplify the algebra involved in the Fourier series. The Cosine and Sine terms are real but become the real and imaginary parts of an imaginary exponential function. Equation (3.1) can then be expressed as

$$f(x) = \frac{a_o}{2} + \sum a_n exp(ink_o)$$
(3.6)

$$\frac{a_o}{2} + \sum_1^\infty A_n(Cos(nk_ox)) + \sum_1^\infty B_n(Sin(nk_ox))$$

$$= \frac{a_o}{2} + \sum(A_n(Cos(nk_ox)) + i(Sin(nk_ox)))$$

$$a_n = A_n \; and \; ia_n = B_n$$

This is an oversimplified case assuming that the terms are independent, which is not true. If we carry out the summations from $-\infty$ to $+\infty$, we get two independent coefficients a_n $(+\infty)$ and a_{-n} $(-\infty)$, then

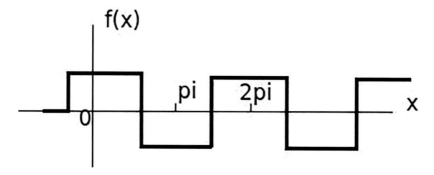

Fig. 3.2 A square-wave pulse

$$a_n + a_{-n} = A_n; \qquad i(a_n - a_{-n}) = B_n$$
further
$$a_n = \tfrac{1}{2}(A_n - i B_n) = \tfrac{1}{2}C_n exp(i\alpha_n)$$
$$a_n = \tfrac{1}{2}(A_n + i B_n) = \tfrac{1}{2}C_n exp(-i\alpha_n)$$
$$i^2(a_n - a_{-n}) = i B_n \qquad a_n + a_n + i B_n = A_n$$
$$i.e. - a_n + a_{-n} = i B_n \qquad a_n = \tfrac{1}{2}(A_n - i B_n)$$
$$a_{-n} = i B_n + a_n$$

Therefore, the Fourier series can be expressed as

$$(x) = \sum_{\infty}^{\infty} a_n exp(ink_o x) \tag{3.7}$$

in exponential form and $a_o = \frac{A_o}{2}$

If the function is itself complex, then A_n and B_n themselves would be complex and otherwise real. Therefore a_n and a_{-n} are complex conjugates i.e. $a_n = a_{-n}^*$.

Example of a square-wave pulse

Let us consider a square wave as our function, as given in Fig. 3.2, to be represented as a Fourier series. Such square functions are encountered in metallurgical context, such as a concentration wave of a solute, in 1-D, in an aligned rod-like or lamellar phase in a eutectic mixture.

It has a constant value over half of its period i.e. $\frac{-\pi}{2}$ to $\frac{\pi}{2}$ and of opposite sign in the other half $\frac{\pi}{2}$ to $\frac{3\pi}{2}$. The function, as defined is an even function and therefore a_n is real. We have discussed these parity conditions above.

$$f(x) = 1 \text{ for } -\frac{\pi}{2} \le x \le \frac{\pi}{2} \text{ and } f(x) = -1 \text{ for } -\frac{\pi}{2} \le x \le \frac{3\pi}{2}.$$

Fig. 3.3 Fourier coefficients as a function of n or k

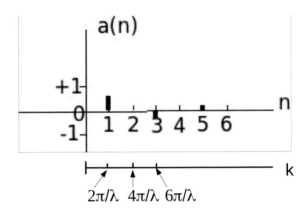

We have $a_n = \dfrac{1}{2\pi} \int_{-\pi}^{\pi} f(x)exp(-inx)dx$

$= \dfrac{1}{2\pi} \int_{\frac{-\pi}{2}}^{\frac{\pi}{2}} exp(-inx)dx - \dfrac{1}{2\pi} \int_{\frac{\pi}{2}}^{\frac{3\pi}{2}} exp(-inx)dx$

$= \dfrac{1}{n\pi} Sin[\dfrac{n\pi}{2}(1 - exp(-inx))]$

Thus, we have $a_o = 0$, $a_1 = \frac{2}{\pi}$, $a_2 = 0$, $a_3 = \frac{-2}{3\pi}$ etc.. We can plot a_n as a function of n, as shown in Fig. 3.3, although n takes integer values, the function can be treated as zero for non-integer values of n (Lipson 1969). Strictly speaking this is not true universally, as in the case of a quasiperiodic function where n takes certain fractional value, a_n is finite.

The function can also be plotted against **k** space. Now that we have all the coefficients, we can sum the series in Eq. (3.1) to get back the square-wave function. The wavelength λ of the function remains unknown though. This information can be obtained from the same integral by replacing the k_o which is equal to $\frac{2\pi}{\lambda}$) by **k** where **k** is $\frac{\lambda}{n}$ corresponding to a harmonic of the wavelength.

i.e. $a(\mathbf{k}) = \dfrac{1}{\lambda} \int_{0}^{one\ wave\ length} f(x)exp(-ikx)dx$

One can readily find the effect of wavelength by taking the new wavelength as a fraction of λ (say $\frac{\lambda}{n} = 0.25$) or an integral multiple of λ(say $\frac{\lambda}{n} = 4$). The coefficients will be either sparsely spaced (for$\frac{\lambda}{n} = 0.25$)or closely spaced (for $\frac{\lambda}{n} = 4$). One important observation from above is that the coordinate x maps the function f(x) in real space while **k** maps the coefficients in reciprocal space (i.e. dimensions are x^{-1}). This has a great bearing on understanding the electron diffraction patterns in Chap. 5.

3.2 Fourier Transforms

Often we deal, in engineering and to some extent in Sciences as well, with non-periodic functions which may be describing objects or phenomena in 1-, 2- or 3-D, i.e. the original function is defined only in a short interval, but not periodic like the earlier discussed function. Fourier series, either Cosine and Sine or exponential functions are periodic in x and thus need to verify whether they can also represent non-periodic functions. We already expressed f(x) in exponential form earlier in Eq. (3.7) as

$$f(x) = \int_{-\infty}^{\infty} (C_n) e^{-2\pi in\frac{x}{\lambda}} dk_n \text{ where}$$

$$C_n = \frac{1}{\lambda} \int_{-\frac{\lambda}{2}}^{\frac{\lambda}{2}} f(x) e^{-2\pi in\frac{x}{\lambda}} dx$$

We can define a K_n vector that represents a wave vector for the successive n values in our Fourier series.

Therefore $K_n = \dfrac{2\pi n}{\lambda}$.

The difference between the successive K_n is dK_n that cannot be any other than $\dfrac{2\pi n}{\lambda}$, n's being integers.

$$dK_n = \frac{2\pi (dn)}{\lambda}$$

where dn = 1.

Since we are interested in a series where $\lambda \longrightarrow \infty$, dK_n is going to be very small. Therefore, K_n can be treated as a continuous function or variable. Thus,

$$f(x) = \int_{-\infty}^{\infty} (C_n \frac{\lambda}{2\pi}) e^{ik_n x} dk_n$$

$$= \int_{-\infty}^{\infty} C(k_n) e^{ik_n x} dk_n \text{ where } C(k_n) = (C_n \frac{\lambda}{2\pi})$$

and $C(k_n) = \dfrac{\lambda}{2\pi} \dfrac{1}{\lambda} \int_{-\infty}^{\infty} f(x)e^{ik_n x} dx$ or simply,

$$f(x) = \int_{-\infty}^{\infty} C(k)e^{ikx} dk \; where \; C(k) = \dfrac{1}{2\pi} \int_{-\infty}^{\infty} f(x)e^{-ikx} dx \qquad (3.8)$$

This is an important result, where C(k) is the Fourier transform of f(x) and *vice versa*. In other words we are saying that f(x) and C(k) are Fourier transforms of each other. But one notices an additional term of 2Π in C(k). So the convention is, C(k)is the *Fourier Transform* of f(x) while *inverse Fourier Transform* of C(k) is f(x). It would also be realised that the non-periodic function f(x) itself is represented by a series and its Fourier Transform also is a series of Fourier coefficients. By an analysis similar to what has been done earlier in Eqs. ((3.3) to (3.4)),

we can easily deduce the following parity conditions:
If f(x) is even and real, then C(k) is even and real.
If f(x) is even and imaginary, then C(k) even and imaginary.
If f(x) is odd and real, then C(k) odd and imaginary.
If f(x) is odd and imaginary, then C(k) odd and real.

Example
Let us get the Fourier transform of a 2-D rectangular aperture of the dimensions given as; length = a and width = b. See Fig. 3.4

$$f(x,y)= \int_{-\frac{a}{2}}^{\frac{a}{2}max} \exp(-iux)dx \int_{-\frac{b}{2}}^{\frac{b}{2}max} \exp(-ivy)dy$$

It is possible to write it as a product of the two functions provided f(x) is independent of f(y), otherwise it should be written as double integral. The corresponding Fourier transform C(k) would be a product of two Sine functions.

$$C(k) = ab \dfrac{Sin(\pi au)}{\pi au} \dfrac{Sin(\pi bv)}{\pi bv}$$

Let a = 10 units and b = 6 units and au = $m_1\lambda$ and bv = $m_2\lambda$, where λ be unity.

C(k) the Fourier transform can be computed for different integer values of m_1 and m_2 as depicted in Fig. 3.5.

Here the central peak is bound by lines given by $m_1 = \pm 1$ and $m_2 = \pm 1$, and so will other subsidiary maxima by higher integer values of m_1 and m_2. The reciprocal relation between (x,y) and (u,v) can be seen in this transform. Further the transform only sees the two opposite boundaries along any direction of the non-periodic object, rectangular aperture in this case. Thus the corner points along the diagonals do not have any effect on the transform. For the same reason, the Fourier transform of a 3-D object(non-periodic 3-D function) needs to be computed in a different way, since one of the sets of parallel boundaries is into the depth and doesn't get represented.

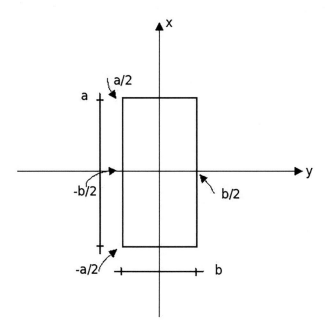

Fig. 3.4 2-D rectangular aperture

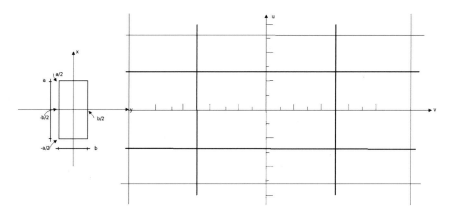

Fig. 3.5 Fourier transform of a rectangular aperture

Recommended reading

1. S.G.Lipson and H.Lipson,"Optical Physics", Cambridge University Press, 1969 or later Editions.
2. Chapter 3 on Fourier analysis, Copyright 2009 by David Morin, morin@physics.harvard.edu (Version 1, November 28, 2009) This file contains the Fourier analysis chapter of a potential book on Waves.

Exercises

Q1. What is the difference between Fourier analysis and Fourier Transformation. Explain why the latter is useful in interpreting electron diffraction patterns.
Q2. Compute the Fourier Transform of a plate-shaped precipitate that is 100 nm long, 10 nm wide and 1 nm thick.

Chapter 4
Transmission Electron Microscope

In this chapter we discuss the essential components of a transmission electron microscope. The components can be divided into (i) illumination system, (ii) image forming system and (iii) viewing and recording system. These are elaborated sequentially in later sections. We will elaborate on the source of electrons, i.e. the electron gun, their different types and the condenser lens system with associated apertures under 'illumination system'. The objective lens forms the most important component. Together with the diffraction and two projector lenses (usually), the objective lens constitutes the image forming system. The objective lens design, its functioning in conjunction with diffraction lens or projector lenses will be part of the Section on 'image forming system'. We will also understand on the basis of respective ray diagrams how a bright field image or a dark field image or a diffraction pattern is formed by the objective lens from the same chosen area of the specimen, in this Section. The electron image (or diffraction pattern) can only be viewed on a fluorescent screen that converts it to an optical image. 'Viewing and recording system' consists of these components together with a recording camera which is either optical or digital. We will discuss the type of fluorescent screen, its persistence, resolution of the digital camera and how to adjust its γ curve in this section.

4.1 Constituents of a Modern TEM

A modern transmission electron microscope can be considered to constitute (i) an illumination system, (ii) an image forming system (iii) a viewing and recording system and (iv) support systems for the above, such as power supplies, vacuum system and so on. A photograph of such a microscope is represented in Fig. 4.1.

© The Author(s), under exclusive license to Springer Nature Singapore Pte Ltd. 2022   25
G. V. S. Sastry, *Microstructural Characterisation Techniques*, Indian Institute of Metals Series, https://doi.org/10.1007/978-981-19-3509-1_4

Fig. 4.1   A transmission electron microscope (model: Tecnai 20G^2)

Here, the first three components are integrated into a column. The support systems are behind the main body of the microscope either in the same room or in an adjoining room. The main body is erected in an exclusive room on a specially built concrete block that lies beneath the floor level and is isolated from the rest of the floor of the room to prevent ground vibrations from reaching the column. Within the instrument the column is further isolated from vibrations by suspending it on air cushions. The room itself is surveyed for any magnetic and acoustic interference prior to the installation and preventive measures are taken generally. Strong air drafts or turbulent air flows are also avoided inside the room. These stringent specifications are laid out in order to achieve the highest resolutions that the microscope is capable of. This will be further discussed in Chap. 7. The ambiance in the room is constantly maintained at 18 °C and 50% relative humidity.

4.2   Illumination System

(i) The illumination system consists of an electron source, called electron gun and two-condenser lenses in tandem.

4.2.1 Electron Gun

Thermionic-emission guns: The basic principle of operation underlying different types of electron guns is the electrostatic force exerted on an electron emitted from the cathode. Under the influence of this force, the electrons get accelerated towards anode. The source of electrons is a sharply bent tungsten wire that is mounted on two electrodes as shown in Fig. 4.2.

The entire electrode assembly is mounted in a ceramic insulator of high dielectric constant. Additionally gas insulators such as sulphur hexafluoride SF_6 are also used in the gun chamber to prevent arcing. The tungsten filament is kept at a high negative potential typically 200 kV relative to an annular disc anode that is kept at earth potential. The filament is also made part of another circuit to pass a small current through, so that it can be heated to high temperature by its Ohmic resistance (Usually the gun and tungsten filament are provided as a standard configuration). The current density of the thermionic emitter can be calculated using the Richardson equation.

$$J_c = AT^2 exp\frac{(-E_w)}{KT}$$

where A is a material constant of the filament $[Am^{-2}K^2]$, T is the absolute temperature of emission[K], E_w is the work function of the filament material (For tungsten it is 4.5 eV), which expresses the energy required to eject an electron from the surface of the filament in vacuum and k is the Boltzmann constant $[eVK^{-1}]$.

Area of the bent tungsten filament tip can be considered to be approximately 100 μm $\times$ 150 μm. Upon increasing the current through the filament, emission current or

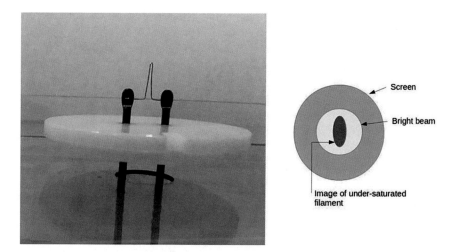

Fig. 4.2 Photograph of a tungsten filament and its schematic image in under-saturated condition

beam current increases and saturates at a certain value of I_f, beyond which we find no further increase in the number of electrons emitted. The central dark schematic image of the filament in under-saturated condition ($I < I_f$), shown as an inset in Fig. 4.2, can also be used to align the gun (filament tip, i.e. source of illumination) along the optic axis of the microscope for maximum brightness. The beam current is the flux of electrons that finally exits the annular anode aperture. The beam is intense in this case, broad in diameter but lacks coherency. At present gun optics has assumed a greater importance owing to the evolution of transmission electron microscope into a multi-functional equipment of highest precision and accuracy and thus many other electron guns are now in use.

In other varieties of guns, a single crystal of lanthanum hexaboride, LaB_6 with a sharp tip oriented in a specific crystallographic direction is mounted on the tip of a tungsten filament. The workfunction of this highly stable ceramic is only 2.4 eV compared to 4.5 eV of tungsten and hence yields more number of electrons compared to tungsten at the same temperature. The crystal tip is about 1 μm in diameter which is responsible for higher brightness of the filament(defined as current per unit area of the emitter per unit solid angle), though the full brightness on the observation screen is lower. It has certain limitations with reference to the required vacuum levels ($>10^{-7}$ torr) and requirement of slow heating and cooling cycles both during filament saturation and switch off.

Field-emission guns: The strong electrostatic field that is created, although being greater than the workfunction of the metal tungsten, is insufficient to extract greater number of electrons even with LaB_6. In another design, called field-emission gun, it is possible to achieve sufficient emission of electrons under the influence of electrostatic field alone by making an atomistically sharp tip at the end of a small tungsten wire that is mounted on the tip of the tungsten filament (Note: this is in place of a LaB_6), Fig. 4.3.

This method is called cold-field emission. These guns give a highly coherent electron beam albeit with low current density, that suits well for Electron Energy Loss Spectroscopy (EELS)(a high resolution in-situ chemical analysis technique) and High Resolution Electron Microscopy (HREM). On the whole, a low beam-

Fig. 4.3 Cold-field emission gun with tungsten tip [after Reimer and Kohl, Transmission Electron Microscopy: Physics of Image Formation, 2nd Edition, Springer Science+Business Media, LLC,2008. With permission]

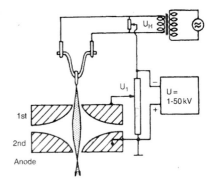

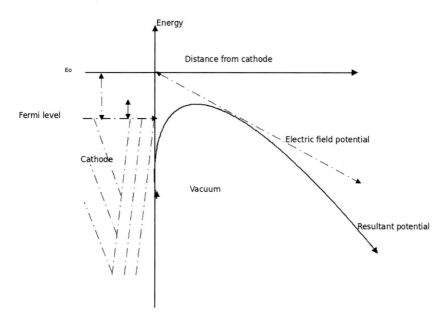

Fig. 4.4 Energy barriers at cold-field emission tip

current renders images that are of low visibility on the fluorescent screen. This low intensity necessitates high sensitivity and high resolution CCD cameras for image capturing and record. Field emission guns can be further classified as (i) cold-field emitters, (ii) Thermal field emitters and (iii) Schottky field emitters.

Cold-field emitters Emission is achieved simply under the influence of potential gradient applied in this type of guns. In order to increase the field intensity, the tip of a tungsten wire is etched down to a radius of less than 100 nm by ion beam thinning. When the electric field on such a tip reaches about 10 V/nm the electrons from the tungsten tip are able to tunnel through the barrier as shown in Fig. 4.4.

Although the overall current is low, a high current density is achieved owing to the small area of the source. These current densities are typically around $10^9 A/m^2 (1 - 3 \times 10^4) A/m^2$ for thermionic emitters) and can be worked out using a simplified Fowler-Nordheim equation:

$$J = a\phi^{-1} F^2 exp[\frac{-v(f)b\phi^{\frac{3}{2}}}{F}]$$

where

$$a = 1.541434 x 10^{-6} \, AeV \, V^{-2}$$

$$b = 6,830890eV^{\frac{-3}{2}} \, V(nm)^{-2}$$

$$v(f) = 1 - f + \tfrac{1}{6} + \ln f$$

$$f = \frac{F}{F_\phi} = (\frac{e^3}{4\pi\epsilon_o})\frac{F}{\phi^2} = 1.439964(\frac{F}{\phi^2})$$

where F_ϕ is the field necessary to reduce the barrier (Sch-Nord) of magnitude equal to workfunction phi(w).

f is the scaled barrier field for SCH-Nord barrier.

a, b are constants.

Brightness in this case is therefore 100x more compared to thermionic emission. A very high level of vacuum is necessary to maintain the tip clean. At a cold start the filament tip needs to be flash-heated several times to clear it from any settled contaminants on the surface. In the process it may have a reduced life. Nevertheless, the overall life of the filament is much more because the replacements are rare and the illumination system remains stable and steady, since no alignments are required in between. The small dimensions of the beam (3 nm) offer very many possibilities for application.

Thermal field-emitters: They function in a way similar to the cold-field emission guns, but are operated at higher temperatures. It ensures cleaning and sharpening of the filament while in use.

Schottky field-emitters: In this type of field emitters, shown in Fig. 4.5, the applied electric field decreases the workfunction of the emitter material. The tungsten tip surface is additionally coated with ZrO_2 which has a lower workfunction compared to tungsten. This combination essentially works like cold-field emitter by incorporating an additional grid current to filter off or retard stray electrons coming from other parts of the filament due to heating. Even though it may function as a thermionic emitter, the brightness and current density of Schottky field emitter is comparable to that of cold-field emitter. Latest developments in this direction are towards using carbon nanotubes as ideal finest field emitters.

4.2.2 Condenser Lens System

All the lenses from this stage of the column are magnetic lenses of the type described earlier. A two-condenser lens system is generally used, see Fig. 4.6. The first condenser, called C1, takes the crossover as the object and forms a highly demagnified image in the forward focal plane. In practice it is often required to change the spot size to different enlargements from the minimum possible. It should be kept in mind

Fig. 4.5 A typical Schottky field emission gun

that there is always a minimum achievable spot size as the electrons undergo mutual repulsion below that size. A change in spot size is achieved by changing the current passing through the coil of the electromagnet that constitutes the pole pieces. A demagnified image of the crossover is formed, which then acts as object for the second condenser C2. It further magnifies the spot and also renders the rays as parallel as possible. The circular fluorescent screen then gets completely covered by the beam. Both C1 and C2 lenses have variable strength to obtain variable spot sizes. Additionally two sets of apertures are provided at C1 and C2 to prevent any divergent electrons from joining the beam.

4.3 Image Forming System

The most important lens of the image forming system of lenses is the objective lens. It is followed by three projector lenses down the optic axis or column of the microscope. The first projector lens is also called intermediate lens or diffraction lens. The objective lens occupies prime position in the design of any microscope, optical or electron, because it is the first lens to form the image of a specimen that is enlarged further by the projector lenses to form final image on the screen. Thus resolution offered by the microscope is decided by the power and performance of the objective lens. The objective lens body also provides space for mounting ancillary equipment such as energy dispersive X-ray analyser which is used to acquire in-situ chemical composition of the particular area of the specimen or particular feature within it, simultaneously. Electron beam which is incident on a thin specimen will be transmitted mostly, but also will be diffracted partly at Bragg angles appropriate to the crystal structure of the specimen. These rays are collected by the objective lens and bring them to focus at its back focal plane as shown in Fig. 4.7.

When these rays travel further down the optic axis, they once again converge and interfere with each other to form the image. Formation of image in an electron microscope differs from that of optical microscope in many ways. The ray paths are, though represented by straight lines in the diagram, strictly helical, and linear representation comes about by a change of coordinate system while representing in the diagram. Therefore, the image orientation with respect to object orientation is at any angle θ that is dependent on the magnification. The specimen is immersed in the magnetic field that comprises the objective lens. In optical microscopes if object is located within the forward focal plane of the convex lens, then the image is a virtual image on the forward side of the lens. Strangely, this interpretation cannot be used in the case of electrons, possibly due to the quantum mechanical behaviour

Fig. 4.6 Two-condenser lens system

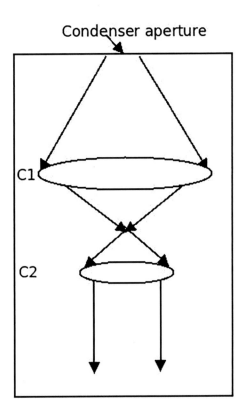

of electrons. The diffraction is not realised in the case of an optical microscope, though it is possible to do so. In the case of electron microscope diffraction takes place from a set of crystallographic planes for which Bragg's law is satisfied and not from the object as a whole (It is customary to diagrammatically represent the pattern as arising from the whole object). In the case of light microscope it is the Fourier transform of the object. The wavelength of light is much larger compared to the interplanar spacing and Bragg's law becomes inapplicable. Further the image plane of the objective lens in an electron microscope is sufficiently far off from the back focal plane or diffraction plane. This gives scope for suitably placing the diffraction lens/ first projector lens below the objective lens. In all current generation TEMs there are three post-objective lenses and first one of the three is exclusively used for displaying the diffraction pattern. When we energise the diffraction lens, the diffraction pattern formed at the back focal plane of the objective becomes its object and it renders an image (i.e. the diffraction pattern itself) on the observation screen, with some magnification (Note: There is no focusing action by this lens). When we wish to observe the image, the diffraction lens is de-energised and the Projector lenses are focused on to the Gaussian image plane of the objective lens and together they render a magnified image on the observation screen. This ability to switch easily between diffraction patterns and images of a chosen area of the specimen is

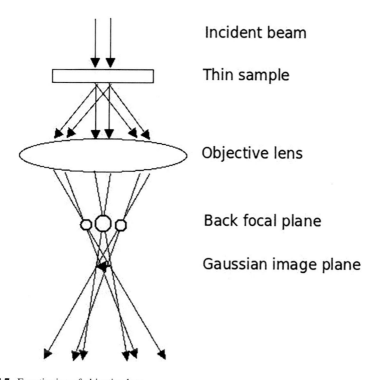

Fig. 4.7 Functioning of objective lens

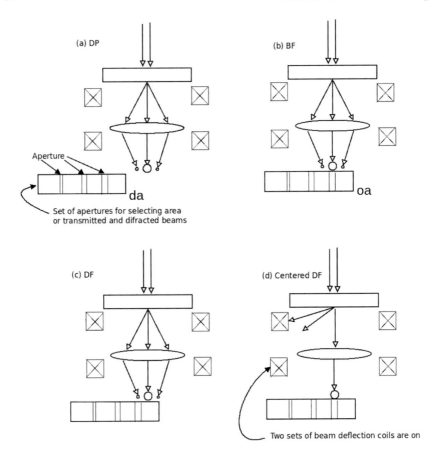

Fig. 4.8 Ray diagrams of diffraction pattern (DP), bright field image (BF), dark field image (DF) and centred dark field image (CDF)

the key feature of the transmission electron microscope. The image and diffraction pattern are correlated and open up several possibilities for materials characterisation [Besides this the large room available at the objective lens also makes it possible to acquire X-rays emitted by the sample surface due to the impingement of high energy electrons for an in-situ chemical analysis]. The corresponding ray diagrams are given in Fig. 4.8.

In (a) a typical diffraction pattern is shown where the diffracted and transmitted beams leaving the bottom surface of the sample are brought to focus at the back focal plane of the objective lens. In actual microscopy an image of the diffraction aperture (da) is brought to coincide with the object plane so as to select the area of the specimen from which the diffraction pattern needs to be obtained. Now an objective aperture (oa) can be inserted in the path of the beams so as to select only the transmitted beam for image formation as shown in (b). An image formed in this way will have the field of view very bright with varying degrees of greyness (See the bottom of the Fig. 4.8b). The greyness indicates that electrons from these regions of the specimen are deflected away from the parallel un-deviated path of the transmitted beam along the optic axis but not thickness variation in the specimen (Will be discussed further in Chap. 7). The darkest regions usually correspond to those orientations which are significantly diffracting the incident electrons. This image is called a Bright field image (BF). Both diffraction and objective apertures are not single apertures but a set of them having different sizes and are mounted in an aperture holder. Instead of the transmitted beam, if any of the diffracted beams, either singly or severally included in the objective aperture, the final image will have a dark field of view in which features that are responsible for those diffracted beams, would appear bright (see Fig. 4.8c). Such image is called darkfield image (DF). In case of specimens with very large lattice constants or in case the chosen diffraction aperture happens to be large, some of the diffracted beams also get included in the objective aperture while taking a BF image. Nevertheless, the field of view remains bright even in such a case. In order to maintain axial symmetry of the image, thus avoiding any distortions, the whole set of beams is shifted so that the desired diffracted beam then coincides with the optic axis of the microscope as shown in Fig. 4.8d. The image then is called a centred dark field image (CDF) which is the default practice in all modern microscopes. As far as selection of the diffracted spot is concerned, we bring a -$\mathbf{g}$ beam into the centre when we desire to get a DF image from a $\mathbf{g}$ vector from a systematics row. Otherwise, if you try to bring the particular $\mathbf{g}$ vector itself to the centre by the tilt knobs provided in DF mode, the intensity of that $\mathbf{g}$beam goes down drastically, Therefore, it is a good practice to use $-\mathbf{g}$ beam for CDF image. For further clarity you may draw the diffracting conditions of formation of systematics row with the help of Ewald Sphere (Refer to Chap. 5). This set of BF, DP and DF comes from the same region of the specimen and is therefore very helpful in characterising the phases present in a specimen by repeating this set from different orientations of the specimen.

It is pertinent to see some details of the specimen holder at this stage. As mentioned in an earlier section, the specimen holder and hence the specimen is placed in the magnetic field of the objective lens and can be rotated on its own axis (α) by default. These are called single-tilt holders. In special holders, a second tilt is also provided (β) whose axis of rotation is perpendicular to the tilt of α. Tilt on the axis of the holder is much wider ($\alpha = \pm 45°$, typical), while the perpendicular tilt is provided generally to a limited extent($\beta = \pm 20°$), which is governed by the gap in the pole pieces. In addition x,y,z translational movements of the holder are also provided by the side entry goniometer. Top entry goniometers are outdated despite their advantages of symmetric location around the optic axis and almost zero drift. Side entry goniometers or holders suffer from severe drawback of specimen drift which can be a problem while performing microscopy at the highest resolution possible. Though the specimen holder is at the centre of the magnetic field of the objective lens, it is required to bring the specimen itself to this exact centre of the maximum field by adjusting the z height of the specimen holder to the eucentric plane. Such adjustment can be realised by observing the image which automatically be in sharp focus (without the need for changing the objective lens current). This height adjustment further helps in performing eucentric tilts of the specimen (The image remains centred while the orientation of the specimen with respect to the beam is changing.). In current generation of microscopes all tilts and translations are motorised. The double-tilt holder and rotation holder are capable of covering a large angular range of reciprocal space of the crystal being investigated. Other special holders such as a heating holder, cryo holder or Faraday-cage holder for magnetic specimens are also available.

4.4   Viewing and Recording

The projector lenses do the job of beam transportation to form the final image on a fluorescent screen. Zinc oxide is the generally used fluorescent material that is coated in powder form on a stainless steel disc. The disc is earthed to conduct away the current generated and also to act as a current measuring device (i.e. intensity of the beam) for deciding the exposure time for recording the image (auto exposure time). Till a decade ago, special grade, low speed diapositive films were used for recording the electron images. Currently, digital image capture and recording are adopted in all microscopes with rapid developments in c-mos technology. For this purpose, there is a digital camera fitted beneath the fluorescent screen. Upon lifting the observation screen, the image forms on the digital camera whose time of exposure and γ curve are set before recording the images. The digital camera transfers the image on to a display monitor.

4.4.1   Digital Image Recording

Digital images were created from a photographic film with the help of an analog to digital converting device in earlier times. With the rapid development in electronics, complimentary metal-oxide semiconductors (CMOS) and charge-coupled devices (CCD) have come into market. Currently, all models of TEMs, SEMs and even optical microscopes are equipped with these devices. A digital camera has become a default option in modern microscopes as photographic films are out of supply and so do the darkroom equipment to develop the film.

CCD Camera: It acquires the image, stores the pixel (picture element), equivalent to a particle of the photographic emulsion, information and transfers the data for subsequent image processing. It has an additional advantage of immediate viewing of the recorded image so that any necessary corrections can be implemented by reverting to the image on the viewing screen of the microscope before changing the area of observation. A number of pixels, which are square in shape, are arranged as 2-D square arrays typically 256×256 or 2048×2048 (some models offer even higher, 4096×4096). The grey levels of image that they can represent are 2^8 (256) or

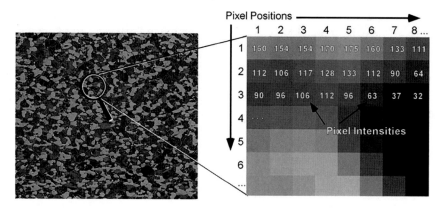

Fig. 4.9 Image to pixel read-out.[after Grande 2012; Metallogr. Microstruct. Anal. (2012) 1:227–243 Reprinted from "Practical Guide to Image Analysis", ASM International, Materials Park, OH (2000). With permission]

$2^{12}(4096)$ and are designated as 8bit or 12bit cameras respectively. The user therefore chooses the best configuration keeping in view the need and cost factor. A typical array of pixels can be represented as in Fig. 4.9 with number of counts or intensity stored in each pixel. The front component of a CCD camera is either a phosphor or yttrium aluminium garnet YAG scintillator which converts the incoming electron of the image into a light photon. The second component of the camera, i.e. the fibre-optic coupling carries forward the photon to the main component of the camera, the charge-coupled device. There the signal is stored as a number of photoelectrons in a pixel. Such a divided assembly prevents the high energy electrons of the beam from reaching the doped silicon wafer of the CCD. The camera is further protected from the heat generated by the high energy electrons by providing Peltier cooling device and circulating water to maintain the temperature around $-32-40\,^{\circ}$C. This prevents thermal noise and resulting dark-current. Too low a temperature again increases the dark-current due to the cooling devices themselves. The currently available cameras offer 10kx10k pixels with high read-out speeds and reduced pixel sizes. These specifications are fast changing.

CMOS Camera: There are, once again two types, one with a direct read-out from each pixel called, MediPIX chip used in medical imaging, the other called monolothic active pixel (MAPS) camera. MAPS CMOS is used in conjunction with TEM. The camera consists of N- and P-type doped semiconducting layer with P^{++} below the N-layer sandwiching a P-epilayer and an array of capacitors and other devices. An electron scattered by the specimen enters the device and creates electron-hole pairs in the P-layer, which discharges the capacitors. The signal is read-out frame by frame, so rapidly that it is almost continuous and the frames are stored. Thus, the CMOS camera gives a very large volume of data and thus necessitates the use of such direct acquisition type camera for special applications.

Performance and efficiency of detectors: Unlike the case of a photographic film camera where the chosen film, e.g. KODAK SO-163 or 4489 has a fixed grain size albeit much smaller than the digital detectors, the performance of digital cameras or detectors is to be evaluated based on certain parameters, before we start using them. These parameters are Detective Quantum Efficiency DQE, which is related to the important parameter of signal to noise ratio s/n, modulation transfer function MTF, gain, dynamic range etc.

DQE is defined for a detector as the detector itself contributes to noise which is already present in the electron image. Therefore, we define the DQE as the ratio of square of the s/n of the output of the detector to the square of the ratio of s/n of the electron beam itself (i.e. inherent in the electron image before the detector). Acceptable values of this ratio range from 0.5 to 1.

MTF, modulation transfer function is the modulus of the Fourier Transform of point spread function of the image, which is obviously much smaller compared to the width of the pixel.

Gain is the ratio of the average number $\bar{I}$, counts delivered by the detector (CCD) to the electrons received per pixel N_e, i.e. $g = \frac{\bar{I}}{N_e}$. An optimum gain only needs to be chosen since maximising the gain of a CCD can result in loss of dynamic range.

Having defined the basic parameters that decide the detector performance, we now discuss the parameters that need to be set while observing and recording a digital image. Note that the image that you see on the fluorescent screen can be very different from that displayed on the computer monitor of your digital detector camera, in terms of its brightness, contrast and the range of grayscale. We should understand what these parameters mean in a digital format. The transformation in grey levels can be due either to (i) the transformation of the grey levels by the processing software to a new level without altering the relative levels amongst the pixels or (ii) transformations that deliberately alter the relative grey levels, an image that has 'balanced' grayscale. The second transformation can be achieved once again in two ways, *viz.*, (i) the histogram method (standard) and (ii) the frequency domain method. This transformation is represented by the so-called gamma curve, which is non-linear. Any transformation that represents the grey levels in a non-linear way such as the gamma curve would render the image with better discernible hues for the eye. The response of human eye to the changes in intensity differences (i.e. contrast) from feature to feature is also *non-linear*.

Histogram method displays the grey level after summing up all the pixels that show a uniform level against the number of pixels. Then the histogram can be equalised using a designed algorithm (in a subsequent section we will show how this is done on the microscope). By this the relative contrast changes while the brightness of the pixel remains the same. Different algorithms can be designed to change the gamma curve in select regions of the image. The second method of frequency domain, a power spectrum of the image is obtained using a Fast Fourier Transform (FFT) to display the periodic structures and their orientations. In fact the FFT facility is extensively used in TEM for checking the beam alignments and also for discerning finer details of HREM images, a job which used to take two days of time earlier.

Fig. 4.10 Adjustment of Gama curve

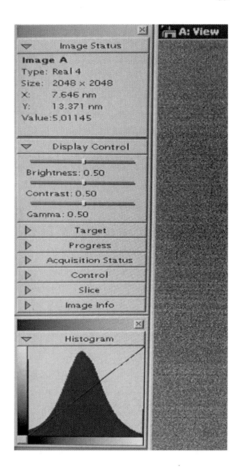

On an actual microscope: Most microscopes these days are equipped with one or the other type of digital camera. The software that controls the camera, besides offering image acquisition, also provides storage and processing capabilities. The display monitor has dropdown menus for choosing different operations. The microscope shown in Fig. 4.1 (installed in the Department) is equipped with a Gatan 2k × 2k camera with Digital Micrograph software. The image is displayed on a 2k × 2k monitor which also displays a dropdown window that shows the histogram of grey levels of the image with brightness, contrast and gamma set at mid-ranges, i.e. 0.5. Now we can alter the histogram by shifting it from linear gamma (see Fig. 4.10) into a curve while observing the changes taking place in the contrast and brightness of the image. We should also keep in mind that the display monitor itself has brightness and contrast controls and also its own gamma.

Storage: The digital image has its own format of storing the processed image called (DM4)(the number 4 represents the version of the software). This file is proprietary and cannot be opened in any other image processing software. Therefore, a more transportable format of Tagged Image File Format (TIFF) is used, which delivers the original data without any compression unlike the more commonly used file formats such as JPEG, and is thus suitable for further processing or printing.

Retrieval: The .tiff files stored on a medium, usually a R/W CD, can be opened in any photo editing software. One should not enlarge these images beyond 2x. Photographic films on the other hand can be enlarged manyfold. DIM files can also be opened in a computer in which a copy of Digital Micrograph software is loaded.

Printing: Printer is yet another device with its own gamma. Therefore, the gamma value of the image should be readjusted and checked on the print iteratively till you get the best results. If the bit value and gamma of the printer are known, then the image can be tuned more easily. The age-old photographic film (Kodak, Ilford or Fuji) has greatest advantage of smaller grain size of the emulsion, about 6μm as against $16–18\mu$m size of pixels of CCD or CMOS cameras. Its storage life also is very large and is not subject to any changes in file formats or drives of digital media. Readers interested in a deeper insight of the digital cameras and digital image processing may refer to (Russ and Neal 2016).

Exercises

Q1. In Chap. 2, we have seen that the image rotates with respect to the object orientation while it is formed by the objective lens at different magnifications. Explain then, why we take help of a diffraction pattern obtained from the same region of observation, in fixing the crystallographic directions of various features such as slip bands and arrays of dislocations.

Chapter 5
Electron Diffraction

Diffraction phenomenon, in general, has been discussed in Chap. 2 earlier. We will now go into details of electron diffraction in the context of a transmission electron microscope and its different modes that are now possible in a modern configuration of the microscope. The theories of electron diffraction have their roots in those that were developed for X-ray diffraction owing to their (X-rays) prior discovery. We will go through the X-ray diffraction principles briefly. These are based on real space arrangement of atoms/planes while electron diffraction is modelled in reciprocal space. In the subsequent sections, we will also discuss concepts of reciprocal lattice, stereographic projection method, representation of electron diffraction in reciprocal space, kinematical theory of electron diffraction, size and shape effects of the diffracting crystal, structure factor and systematic absences. Further, we will also discuss methods of indexing electron diffraction patterns, Kikuchi patterns and applications of electron diffraction analysis such as determination of orientation relationships etc. Effect of ordering of the crystal as well as lack of any order, i.e. amorphous materials) will also be considered in this section. We will devote a separate section for describing the details of convergent beam electron diffraction (CBED), its variants and applications. At the end we will survey other techniques of diffraction briefly.

5.1 Laue Diffraction

Let us consider a linear periodic chain of atoms which act as scatterers when X-rays with a planar wavefront are incident on them. Lets further consider that the propagation of the incident plane wave is represented by its wavevector $\mathbf{k}$, whose wave magnitude is given by the reciprocal of wavelength of the monochromatic X-rays used. Its direction is normal to the wavefront. The periodic chain of atoms

© The Author(s), under exclusive license to Springer Nature Singapore Pte Ltd. 2022   41
G. V. S. Sastry, *Microstructural Characterisation Techniques*, Indian Institute
of Metals Series, https://doi.org/10.1007/978-981-19-3509-1_5

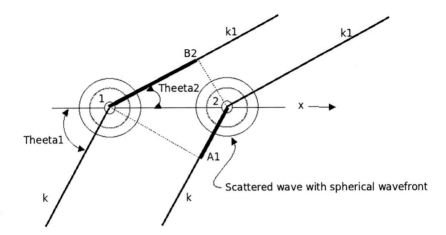

Fig. 5.1 Laue diffraction

we consider to be along X-axis of the periodic crystal with period a_1. Due to the interaction with the incident wave, the periodic set of atoms along the row scatter in all directions forming a spherical wavefront. For interference of the scattered waves to take place, they should arrive with the same phase at a distant point or a screen. Their interference leads to a diffraction pattern. When we project x_1 onto the incident wavevector $\mathbf{k}$, it will lie on the dotted line A1 as shown in Fig. 5.1. For all arbitrary values of x_1, A1 represents the locus of wavefronts that will be in phase. Consider projection of x_1 on to the scattered wavevector $\mathbf{k}_1$ in a similar way. It lies on the dotted line B2, representing again the locus of wavefronts that would be in phase for all arbitrary values of x_1.

Therefore, the optical path-length through the scatterer 2 is shorter than that through 1 and is given by

$$\mathbf{k}_1.\mathbf{x}_1 - \mathbf{k}.\mathbf{x}_1 = (\mathbf{k}_1 - \mathbf{k}).\mathbf{x}_1 \tag{5.1}$$

For maximum reinforcement of the scattered wave, their path difference should be an integral multiple of the wavelength, say h

$$\text{i.e.} (\mathbf{k}_1 - \mathbf{k}).\mathbf{x}_1 = h$$

where 'h' takes integer values 0, 1, 2,... When h = 0, we see that $\mathbf{k}_1 = \mathbf{k}$ and thus complete constructive interference of the waves takes place in the incident wave direction, meaning the transmitted beam. The direction of the diffracted wavevector $\mathbf{k}_1$ in fact is one of the several, describing the surface of a cone. The axis of which is the row of scatterers. This is so because the individual atoms scatter in a spherical way and their path lengths will be the same in all planes containing the row of atoms. Thus there will be several congruent cones of intensity corresponding to successive

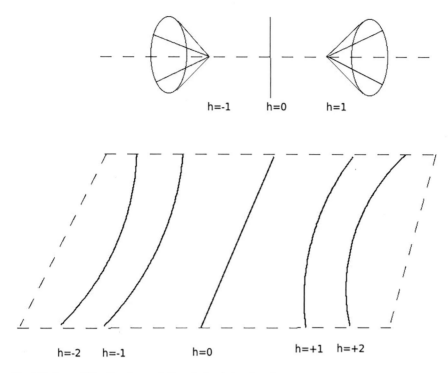

Fig. 5.2 Laue diffraction from a 1-D periodic chain of scatterers

orders/integer values of h (see Fig. 5.2) in the positive direction of $\mathbf{x}_1$ and a similar
number in the negative direction corresponding to –h.

If we keep any observation screen parallel to the string of atoms, then hyperbolic
curves of intensity will be observed corresponding to each cone except the h = 0
case, where it will be a straight line.

If we consider another periodic string of atoms in the direction along the unit
vector $\hat{x}_2$ at an arbitrary angle to $\hat{x}_1$ and with period a_2, we form a periodic grid of
atoms in 2-Dimensions. The atoms in this grid can be located by the vector

$$\mathbf{x}_{(1,2)} = a_1\hat{x}_1 + a_2\hat{x}_2 \tag{5.2}$$

The conditions for complete constructive interference of the scattered waves from
any one atom and its two neighbours from two rows can be derived by satisfying the
following conditions.

$$(\mathbf{k}_1 - \mathbf{k}).\mathbf{a}_1 = h_1 \text{ and } (\mathbf{k}_1 - \mathbf{k}).\mathbf{a}_2 = h_2 \tag{5.3}$$

where h_1 and h_2 are again integers, which can be different from each other (Fig. 5.3).

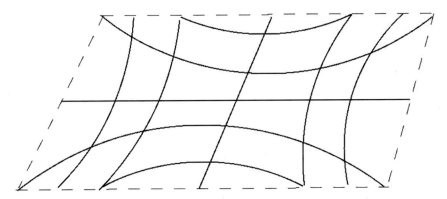

Fig. 5.3 Laue pattern from a 2-D plane of scatterers

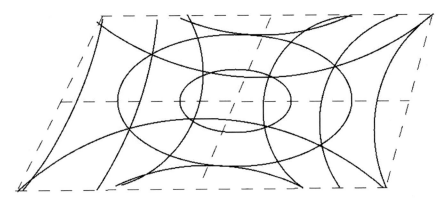

Fig. 5.4 Laue pattern from a 3-D crystal

The spacing of the hyperbolae depends on the relative magnitudes of a_1 and a_2, when they are longer, the lines will be closer. The lines joining the points of intersection of the hyperbolae with the origin indicate the directions along which intense diffracted beam will emerge. Extending these arguments to a third row of atoms along x_3 and at an acute angle (not necessarily 90°) we get concentric ellipses (if x_3 is at 90° they will be circles). On the observational screen held parallel to the 3-D grid of atoms, i.e. 3-D crystal, there will be some common points of intersection between the two sets of hyperbolae and ellipse/circle. The line joining these points of intersections and centre of 3-D grid indicate the directions along which intense diffracted beams can be observed (Fig. 5.4).

We then have a set of three equations which should be satisfied.

$$(\mathbf{k}_1 - \mathbf{k}).\mathbf{a}_1 = h_1; \; (\mathbf{k}_1 - \mathbf{k}).\mathbf{a}_2 = h_2; \; \text{and} \; (\mathbf{k}_1 - \mathbf{k}).\mathbf{a}_3 = h_3 \qquad (5.4)$$

The above are the famous Laue equations for diffraction from a periodic crystal.

5.2 Bragg's Law

Let us also examine how Bragg had derived these diffraction conditions using the crystal as a set of parallel planes. The crystal planes repeat periodically, with period d, in any orientation of the crystal as shown in Fig. 5.5.

The crystal planes consist of periodic set of atoms with different densities. When X-rays are incident at an angle on top of the crystal, i.e. on 1-1 plane, say the ray AOB, they get reflected at the same angle θ as are incident. The ray CPD which is incident on the next crystal plane 2-2 also gets similarly reflected and the process continuous for all the parallel planes up to a certain depth in the crystal (m-m). The incident and reflected rays lie in the same plane with reference to the crystal surface and for any arbitrary angle of incidence θ, the intensity of the reflected rays will be very small as there is no special phase relation amongst them leading to constructive interference. For certain specific angles of incidence, however, the extra optical path travelled by the reflected ray CPD w.r.t. AOB, i.e. KPL happens to be an integral multiple of phase difference or integral number of wavelengths. In such a case the reflected rays OB and PD arrive at a distant observation plane in phase. They interfere constructively at this plane and lead to reinforcement of amplitudes. For the configuration given in Fig. 5.5, this path difference works out to be 2dSinθ. So the beam is said to be diffracted with full intensity under this condition, called Bragg condition.

$$\text{i.e. } 2d\text{Sin}(\theta) = n\lambda \tag{5.5}$$

which is called Bragg's law.

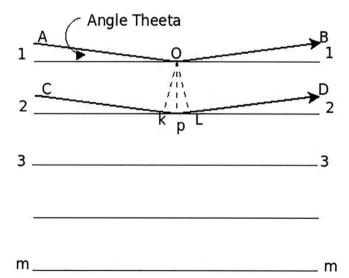

Fig. 5.5 Bragg diffraction

There are certain assumptions implicit in this derivation. Firstly, the treatment is a mixture of both geometrical and physical optics since the top ray is treated geometrically and the other rays according to physical optics. Secondly, refraction of X-rays, as they enter the crystal leading to their deviation from the angle of incidence θ for the subsequent planes, is ignored. Additionally atomic density of the crystal planes is not considered to be a parameter influencing the reflected intensity, e.g. the atomic density of high index planes is low while that of the low index planes is high. Thus, in the case of high index planes, there will be many more atoms belonging to other planes which also interact with the incident plane at that angle.

Both the above treatments are under kinematical conditions, i.e. they consider only the geometrical paths of the rays in certain directions without taking into account the energetics involved in doing so (which forms the dynamical treatment). Bragg's law entirely refers to the physical space while Laue equations involve the concepts of reciprocal space to some extent (the use of wavevector k). You will realise in the course of this topic that complete treatment of electron diffraction in reciprocal space makes many concepts more clear and hence adopted in electron microscopy. Lets first understand the Bravais lattice of a real crystal projected in a reciprocal space.

5.2.1 Reciprocal Lattice

Any general reciprocal lattice is a 3-D periodic set of points that displays the symmetry relations of the corresponding real lattice and bears a reciprocal relation to the corresponding real lattice in its metric. A reciprocal lattice vector can be represented by

$$\mathbf{r}^* = h\mathbf{a}^* + k\mathbf{b}^* + l\mathbf{c}^* \tag{5.6}$$

where $\mathbf{a}^*, \mathbf{b}^*$ and $\mathbf{c}^*$ are the basis vectors of the reciprocal lattice.

This reciprocal lattice can be constructed from the corresponding real lattice in the following manner. Let abc be any real lattice with basis vectors $\mathbf{a}, \mathbf{b}$ and $\mathbf{c}$ and mutual inter-axial angles α, β and γ. Erect a vector $\mathbf{c}^*$ perpendicular to the plane constituted by $\mathbf{a}, \mathbf{b}$. Let the angle that it subtends with $\mathbf{c}$ be δ and its magnitude be such that $\mathbf{c}^* . \mathbf{c} = 1$

i.e. $\mathbf{c}^* . \mathbf{c}\ \mathrm{Cos}\delta = 1$. See Fig. 5.6.

Similarly other reciprocal lattice basis vectors will be related to corresponding real lattice vectors as

$$\mathbf{a}^* . \mathbf{a} = 1 \ \text{ and } \ \mathbf{b}^* . \mathbf{b} = 1$$

Volume of the real lattice unitcell can be worked out as

$$V = (\mathbf{a} x \mathbf{b}) . \mathbf{c} = (\mathbf{b} x \mathbf{c}) . \mathbf{a} = (\mathbf{c} x \mathbf{a}) . \mathbf{b}$$

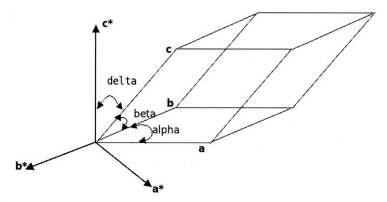

Fig. 5.6 Construction of a reciprocal lattice from real lattice framework

for the reason that area of the base is given by

$$|\mathbf{a}||\mathbf{b}|Sin\alpha$$

and altitude is equal to

$$|\mathbf{c}|Cos\delta$$

Now $\mathbf{c}^*$ can be expressed in terms of the corresponding real lattice as

$$\mathbf{c}^* = \frac{\mathbf{a}x\mathbf{b}}{V} \text{ and on similar lines}$$

$$\mathbf{b}^* = \frac{\mathbf{c}x\mathbf{a}}{V}$$

$$\mathbf{a}^* = \frac{\mathbf{b}x\mathbf{c}}{V}$$

Now that the relationship with corresponding real lattice is established, we can understand many more relations in that context.

Let us find the magnitude of the vector $\mathbf{c}^*$ that we defined from the equation

$$|\mathbf{c}^*| = \frac{abSin\alpha}{(abSin\alpha)(|\mathbf{c}|Cos\delta)}$$

$$= \frac{1}{d_{(001)}}$$

i.e. magnitude of $\mathbf{c}^*$ will be reciprocal of the interplanar spacing of the (001) planes in real lattice and direction being perpendicular to the set of planes. In general any reciprocal lattice vector $\mathbf{r}^*$ will have a magnitude which is reciprocal of the interplanar spacing that it correspond to

$$|\mathbf{r}^*| = \frac{1}{d_{(hkl)}}$$

Therefore, it can be stated that a reciprocal lattice point represents a family of planes of the corresponding real lattice and can thus be correlated with any physical phenomenon such as diffraction taking place from that set of planes. In fact, von Laue's treatment of kinematical diffraction from a crystal utilises these concepts.

In the above approach, we are successful in preserving the interplanar distances in the projection. There is yet another approach in which we can project a crystal in such a way as to preserve the angular relationships of the crystallographic planes disregarding their relative spacings in the crystal. There are many methods in the approach, however, the stereographic projection is the most widely used.

5.2.2 Stereographic Projection

This graphical method of representing the angular relationships amongst various crystal planes of a chosen crystal is one of the many methods of projecting 3-D objects into 2-D graphical representations. We have orthographic projection and isometric projection which preserve length scales of the object in 2-D projections. When we wish to preserve the angular relationships of the object in 3-D we take the help of spherical projection. In this method, a small crystal is assumed to be placed at the centre of a transparent unit sphere. A pole is erected perpendicular to a chosen plane of the cubic crystal, say the top surface (001) plane. This pole intersects the sphere at a point N as shown in Fig. 5.7.

Similarly, the crystallographic direction [001] is represented by a plane perpendicular to it, i.e. the (001) plane and it intersects the sphere in a circle O, which passes through the equator of the sphere. Thus in spherical projection the crystallographic planes are represented by their poles which are perpendicular to them and crystallographic directions are represented by the planes that are erected perpendicular to the directions. The former are represented by points of intersection and the latter are represented by circles of intersection on the spherical surface. Thus, there is a reciprocal relation in the representation of planes and directions, planes are represented by directions and vice versa. All planes which are parallel to the (001) plane, i.e. the (00h) will be represented by a single pole and its point of intersection with the sphere of projection. The $(00\overline{h})$ will be represented by the diametrically opposite point on the sphere. In this way all the crystal planes and directions of interest can be projected, but the projection, which is the sphere, needs to be carried along and the measurement of angles need to be derived from the latitude and longitude positions.

In stereographic projection this difficulty is overcome by doing the projection on to a plane rather than on to a sphere. We proceed in a similar way by locating a tiny crystal, a unitcell, at the centre of the sphere and mark the North and South poles for fixing the frame of reference conveniently, the East and West poles may also be marked on the sphere as shown in Fig. 5.8.

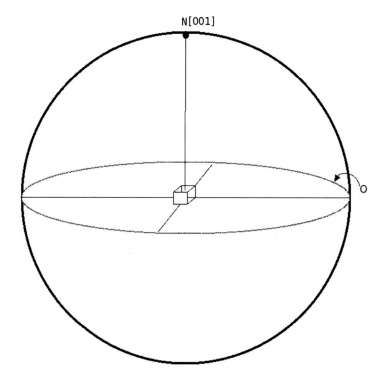

Fig. 5.7 Spherical projection

The tiny crystal is then oriented at the centre such that the [001] direction of the cubic crystal is oriented along ON, i.e. the North direction, the [100] direction along the direction perpendicular to E-W line and the [010] direction along OE, i.e. the East direction. Let OQ be a pole of interest which intersects the sphere at Q. The pole OQ makes an angle ϕ, angle of dip, with the N-S direction and is away by an angle θ from the NESW plane, a plane passing through the N-S and E-W poles and which is taken as a reference plane from where angles of rotation are measured. The equatorial plane WRE is considered the projection plane. The projections are confined within the equatorial circle WRE. When point Q is joined to the South pole S, it intersects the projection plane at point P, which is defined as the stereographic projection of pole Q. Therefore, point P represents the (hkl) plane in the stereograph, to which OQ is the corresponding pole. In fact, pole OQ represents all such planes which are parallel to the (hkl) plane and thus P represents that family {hkl} in the stereograph. All the planes of interest in the crystal can be projected in a similar way through their corresponding poles and generate a 2-D map on the equatorial plane. Such a map is called the stereogram.

It can be easily understood that if a pole intersects the sphere of projection in the lower hemisphere, then the line joining the intersection point to the South pole would not pass through the equatorial plane within the bounds of the equatorial circle

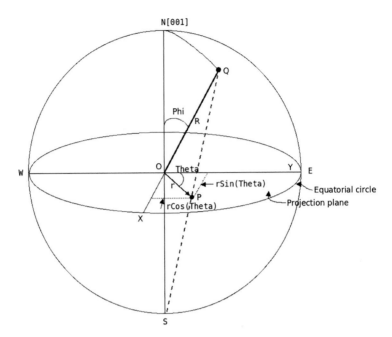

Fig. 5.8 Stereographic projection

but outside of it. If it is essential to represent such a pole in the stereogram for any reason, then its corresponding pole in the upper hemisphere (northern hemisphere, of crystallographically opposite sign) is located and its intersection in the equatorial plane is then represented by a dotted circle. In order to complete the mapping, we need to arrive at the (x, y) coordinates of the point p with respect to the reference frame on the equatorial plane. These can easily be arrived at using the geometrical relations of the projection sphere. On the equatorial plane projection of Q is represented by point p and op is located at an angle θ from the reference plane NESW. According to one convention the intersection of the reference plane is taken as the y-axis on the projection circle and X-axis is defined perpendicular to it as shown in Fig. 5.8.

The (x, y) coordinates of point P can now be worked out as r $\mathrm{Sin}(\theta)$ and r $\mathrm{Cos}(\theta)$, respectively, where r is the radial distance from the Centre O. It can be related to R the radius of the sphere by considering triangles OQP and OSP in Fig. 5.8 as

$$r = R\mathrm{Tan}\frac{\phi}{2}$$

Thus, coordinates of the pole Q, which represents the (hkl) plane in the real lattice, are completely defined in the stereogram. This procedure can be repeated to project all planes of interest for a chosen orientation of the crystal. Crystallographic directions get projected as arcs of great circles on the projection plane. Figure 5.9 shows a standard projection of a cubic crystal along [001] direction.

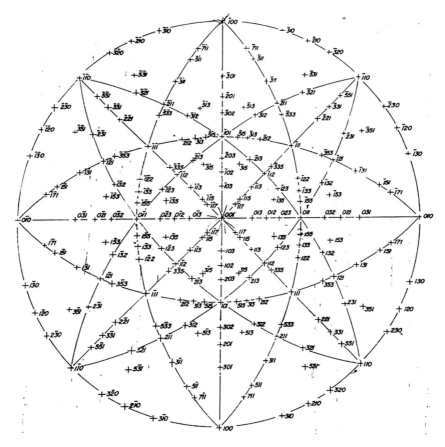

Fig. 5.9 A standard cubic [001] projection: [after O. Johari and G. Thomas, The stereographic Projection and its Applications, InterScience Publishers, a Division of John Wiley & Sons, 1969. With Permission]

As mentioned earlier the poles on the equatorial circle are 90° to the axis of projection, i.e. [001]. They constitute a family of planes that are parallel to the axis of projection [001] and hence are called a zone. The axis of projection then becomes the zone axis for that family. Further, the symmetry of the arcs of the great circles would be reflecting the symmetry of the axis, i.e. four-fold symmetry in this case. Once the stereogram is available for the axis of interest, the angles between any two desired planes can be determined using a measuring tool designed by Pennfield originally and later by Wulff.

The Wulffnet, which is extensively used at present, has a chosen basic circle that is divided into several meridian lines and latitudes at 2° interval. The radii of these arcs of circles can be calculated using the formulae given in Appendix 5.1 to generate a Wulffnet of desired diameter with meridians and latitudes at desired degrees of interval. A standard Wulffnet is reproduced in Fig. 5.10. For measuring

Fig. 5.10 Wulff net at 2°

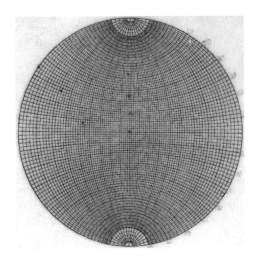

angles between planes, i.e. projected poles, one needs a Wulffnet of the same diameter as that of the chosen stereogram. If A and B are two poles of interest as shown in Fig. 5.11, then the stereogram should be so rotated with respect to the overlaid Wulffnet that the two poles lie on the same meridian line.

Then the difference in latitudes between A and B gives a measure of the angle between the two crystallographic planes. In case, the crystal is not from the isometric class, the (hkl) plane and the corresponding direction, e.g. (101) pole and [101] zone axis do not coincide and therefore get projected at different locations in the corresponding standard stereogram. For further study of stereographic projection, its properties, other types of projections, measuring nets and applications you are suggested to refer to Appendix 5.1 and an exhaustive treatment of the topic by Johari and Thomas (1969).

5.3 Representation of Electron Diffraction in Reciprocal Space

The case of diffraction from a thin specimen in a transmission electron microscope is related more to the transmission Laue diffraction conditions rather than to the Bragg conditions, since the diffracted beams also travel in the same direction as that of the incident and transmitted beams. Let us construct the diffraction conditions completely in reciprocal space. According to the deBroglie principle, the wavelength of electrons under a moderate accelerating voltage of 200 kV is approximately 0.028Å. The wave vector of the incident electrons ($\frac{1}{\lambda}$) will be very large and the locus of the wavevector is a very flat sphere, called Ewald sphere (named after Ewald who developed kinematical theory of diffraction). Let us now consider any common f.c.c.

Fig. 5.11 Measuring the
angle between two poles

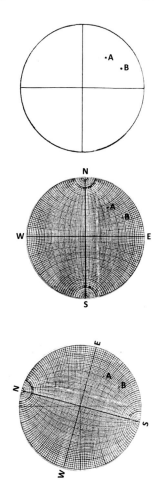

metal such as aluminium with a lattice parameter of 0.40 nm (keeping the accuracy
of electron diffraction in mind) and construct a reciprocal lattice section normal to
[001] direction. The section will be a square grid of points representing 200, 020
and 220 type of planes. In the real crystal these planes have d spacings of 0.203 and
0.143 nm with corresponding Bragg angles of 0.4 and 0.56 degrees for the chosen
electron beam of wavelength of 0.028 Å electron beam. The Ewald sphere, though
strictly according to geometry can only touch one point of the reciprocal lattice grid
(corresponding to the incidence of the electron beam on the crystal surface), can
pass through several points of the reciprocal lattice for a given orientation due to its
flatness, as shown in Fig. 5.12 (similar to a slightly deflated car tyre on a road!). If
the lattice parameters of the crystal and consequently the d spacings are large, then
the reciprocal lattice sections, as well as the inter-layer spacings, will have points

Fig. 5.12 Intersection of
Ewald Sphere with a
reciprocal lattice section(s)

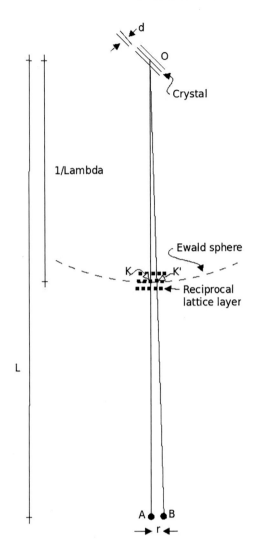

which are very closely spaced. In such crystals, the Ewald sphere can also touch the
reciprocal lattice points from the upper layer as shown in Fig. 5.12.

Yet another factor, which influences the number of reciprocal lattice points sampled by the Ewald sphere is the size of the diffracting crystal. If the thickness of the
crystal is three to four orders of magnitude smaller than its size in other dimensions,
then the reciprocal lattice points corresponding to the planes perpendicular to the thin
direction get elongated in that direction. The reciprocal lattice points now have the
shape of thin rods and are called 'relrods'. A typical electron microscopy specimen is
a few millimetres (maximum 3 mm in diameter) in its length and breadth but is only
a few tens of nanometers in its thickness for rendering it electron transparent. Thus,

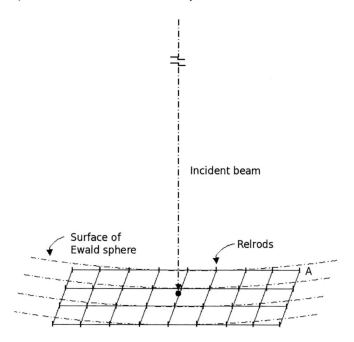

Fig. 5.13 Sampling of many 'relrods' by Ewald's sphere at one direction of incident beam

all the points of the reciprocal lattice section parallel to the surface of the foil show 'relrods' parallel to the smallest dimension of the crystal, i.e. thickness direction as given in Fig. 5.13.

In such a case the Ewald sphere not only passes through many reciprocal lattice points, but also intersects the 'relrod' extensions of many others. Just as the reciprocal lattice is a concept, so is the relrod and there are no physical rods present. It is the shape transform of the small crystal and tells us the intensity distribution at the reciprocal lattice point as seen in the derivation given below.

5.4 Intensity of a Diffracted Beam From a Thin Crystal Under Kinematical Conditions

Let us assume that our specimen is a thin slab of a single crystal kept at $(0, 0)$ tilt position normal to the path of incident electron beam in the microscope. The crystal can be visualised as a decoration of atoms on an imaginary lattice of periodic set of points as explained in the section on Basic Crystallography (Appendix 5.2). Alternatively, it can also be viewed as a set of unitcells associated with the lattice points where the atoms associated with the unitcell scatter the incident beam of electrons. The scattered waves combine together to give rise to diffracted beams

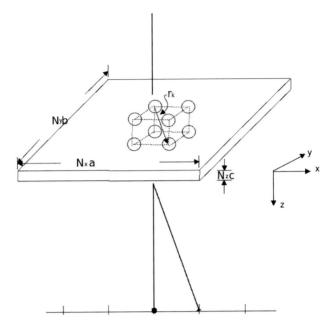

Fig. 5.14 Diffraction by a thin slab of crystal. A typical Unitcell at the point of incidence of the beam is shown enlarged

from different crystal planes. Extending the concepts used in the derivation of Laue equations for kinematical diffraction of X-rays to electron diffraction, we can propose that diffraction takes place from a (hkl) plane when the diffraction vector **k** equals the reciprocal lattice vector **g** of the (hkl) plane.

i.e. $\mathbf{k_o} + \mathbf{k} = \mathbf{k}_d$ or $\mathbf{k_o} + \mathbf{g} = \mathbf{k}_d$ where $\mathbf{k_o}$ and $\mathbf{k}_d$ are the incident and diffracted wavevectors, respectively.

The thin slab of crystal can be considered as a stack of unitcells N_x along $\hat{x}$, N_y along $\hat{y}$ and N_z along $\hat{z}$ directions where $\hat{x}$, $\hat{y}$ and $\hat{z}$ are three mutually perpendicular basis vectors of the crystal lattice. The dimensions of the crystal would then be $N_x a$, $N_y b$, $N_z c$ along x, y and z directions, where a,b and c are the lattice parameters of the crystal as shown in Fig. 5.14.

So the total number of unitcells in the crystal would be $N = N_x N_y N_z$, we can formulate an expression for amplitude of the scattered wave from a unitcell in terms of the amplitude and phase factors of the scattered electrons.

$$\phi_{unitcell} = \frac{e^{2\pi i \mathbf{k} r}}{r} \left(\sum_i f_i(\theta) e^{2\pi i (\mathbf{k}) \cdot (\mathbf{r}_k)} \right) \tag{5.7}$$

where the summation describes the amplitude of scattered wave from the atoms of the unitcell due to the incident wave represented by the constant factor outside the summation, $f_i(\theta)$ is the atomic scattering factor representing the amplitude in the direction of the diffracted beam. When summed over all the atoms of the unitcell Eq. (5.7)

$$\phi_{unitcell} = \frac{e^{2\pi i \mathbf{kr}}}{r}(F(\theta)) \tag{5.8}$$

When we consider all the unitcells in the crystal slab we need to sum the above Eq. (5.8) over all the unitcells to yield,

$$\phi_{unitcell} = \frac{\pi\, ai}{\xi_g}\left(\sum_n e^{-2\pi i(\mathbf{k})\cdot(\mathbf{r}_n)} e^{-2\pi i(\mathbf{k}_D)\cdot(\mathbf{r}_k)}\right) \tag{5.9}$$

where r_n is the positional vector of the unitcell and ξ is the extinction distance, sometimes also known as characteristic length.

$$\xi_g = \frac{\pi V_c Cos\theta_B}{\lambda|F_g|} \tag{5.10}$$

Here, V_c is the volume of the unitcell, θ_B is the Bragg angle, λ is wavelength of the electron beam incident on the crystal and $|F_g|$ is the modulus of structure factor for the particular crystal plane under consideration. In the above Eqs. 5.9 and 5.10, $\mathbf{g}$ represents the reciprocal lattice vector of the set of diffracting crystal planes.

Shape factor: We now rewrite the Eq. (5.7) in terms of number of unitcells along the three mutually perpendicular directions of the crystal. The position vector $\mathbf{r}$ of the scatterer can be written in terms of the lattice vectors and the vectors that represent the atoms within a unitcell, i.e. $\mathbf{r} = \mathbf{r}_g + \mathbf{r}_k$

$$\phi_{\Delta K} = \sum_{r_g}\sum_{r_k} f_{at}(r_k)e^{-2\pi i(\mathbf{k})\cdot(\mathbf{r}_g+\mathbf{r}_k)} \tag{5.11}$$

since f_{at} is same for all the unitcells as they are decorated by the atoms in the same way. The diffraction vector $\mathbf{k}$ and position vector $\mathbf{r}_g$ can be resolved along the basis vectors:

$$\mathbf{k} = k_x\hat{x} + k_y\hat{y} + k_z\hat{z} \text{ and}$$
$$\mathbf{r_g} = ma\hat{x} + nb\hat{y} + pc\hat{z}$$
$$\phi_{\Delta K} = \left(\sum_{r_g} e^{-2\pi i(\Delta\mathbf{k})\cdot(\mathbf{r}_g)}\sum_{r_k} f_{at}\mathbf{r}_k e^{-2\pi i(\Delta\mathbf{k})\cdot(\mathbf{r}_k)}\right) \tag{5.12}$$

$$\text{i.e. } \phi_{\Delta k} = S(\Delta k)F(\Delta k) \tag{5.13}$$

Shape factor $S(\Delta\mathbf{k})$ is given by

$$S(\Delta\mathbf{k}) = \sum_{m=0}^{N_x-1}\sum_{n=0}^{N_y-1}\sum_{p=0}^{N_z-1} e^{-2\pi i(k_xam+k_ybn+k_zcp)} \tag{5.14}$$

$$S(\Delta\mathbf{k}) = \sum_{m=0}^{N_x-1} e^{-2\pi i(k_xam)}\sum_{n=0}^{N_y-1} e^{-2\pi i(k_ybn)}\sum_{p=0}^{N_z-1} e^{-2\pi i(k_zcp)} \tag{5.15}$$

$$S = 1 + v + v^2 + v^3 + \dots + v^{N-1} \quad \text{where } v = e^{-2\pi i(\mathbf{k_x}a)}$$

$$= \frac{1 - v^N}{1 - v}$$

$$S = \frac{1 - e^{-2\pi i(\mathbf{k_x}aN_x)}}{1 - e^{-2\pi i(\mathbf{k_x}a)}}$$

and intensity I is given by

$$S^* \cdot S(K_x) = I = \frac{1 - e^{2\pi i(\mathbf{k_x}aN_x)}}{1 - e^{2\pi i(\mathbf{k_x}a)}} x \frac{1 - e^{-2\pi i(\mathbf{k_x}aN_x)}}{1 - e^{-2\pi i(\mathbf{k_x}a)}} \tag{5.16}$$

where the first term is the complex conjugate of $S(K_x)$ and hence has a sign change. Equation (5.16) is of the form

$$f(x) = \frac{2 - (e^{-i\theta N_x}) - (e^{+i\theta N_x})}{2 - (e^{-i\theta}) - (e^{+i\theta})}$$

From trigonometric relations in terms of exponential function, we know

$$Cos\theta + i\,Sin\theta = e^{i\theta}; \; Cos\theta - i\,Sin\theta = e^{-i\theta}$$

Therefore,

$$S^* \cdot S(K_x) = \frac{2 - 2Cos(2\pi k_x aN_x)}{2 - 2Cos(2\pi k_x a)}$$

$$= \frac{2 - 2[1 - 2Sin^2(\pi k_x aN_x)]}{2 - 2[1 - 2Sin^2(\pi k_x a)]}$$

Since $Cos(2\theta) = 1 - 2Sin^2\theta$

$$S^* \cdot S(K_x) = \frac{Sin^2(\pi k_x aN_x)}{Sin^2(\pi k_x a)} \tag{5.17}$$

This is the expression for intensity of scattered beam down the column of N_x scatterers. When $k_x a$ is an integer the expression in Eq. (5.17) becomes indeterminate.

We can overcome this state by rewriting the expression in Eq. (5.15) in terms of exponential functions.

$$S(\Delta\mathbf{k}) = \sum_{m=0}^{N_x-1} e^{2\pi i(\text{integer})m} \sum_{m'=0}^{N_x-1} e^{-2\pi i(\text{integer})m'}$$

$$= \sum_{m=0}^{N_x-1} 1 x \sum_{m'=0}^{N_x-1} 1$$

$$= N_x^2 \tag{5.18}$$

Therefore intensity of a diffracted beam from the small crystal scales as square of the number of unitcells in that direction (i.e. x). When their number is doubled, say, the intensity is quadrupled. However, every doubling of the number of diffracting planes, the **k** value gets halved and therefore, the intensity varies linearly as N. The peak intensity increases as the number of cells increase and the satellite peaks come closer to the main peak as shown in Fig. 5.15

$$\text{Therefore, I} = \text{Sin}^2(\pi k_x a N) x(k) ...where\ x(k) = \frac{1}{(\pi k_x a)^2},$$

is an envelope function of the subsidiary maxima.

Size effect on reciprocal lattice point and diffraction spot: The expression (5.17) is derived for intensity distribution along z or thickness direction. The same expression can be extended to the case of small cuboid of a crystal. In that case the intensity distribution in the reciprocal lattice will be extended in the three mutually perpendicular directions. One may not be able to see the extensions due to the satellite peaks often. When the Ewald sphere intersects these 'relrod' extensions oriented tangential to it, streaks of intensity are seen in the diffraction pattern along the corresponding crystallographic direction passing through the diffraction spot. These effects are pictorially shown in Table 5.1. If the crystal shape is other than a thin slab, then the above formulation cannot give us a clue to the 'relspace' shape. We need to use Bessel functions to infer the same. If the crystal is spherical, then there will be concentric shells in 'relspace' associated with each reciprocal lattice point and again the diffraction spots will be surrounded by diffuse halos. In the case of crystals that are either thin plates or when large grains have either coherent twin boundaries or stacking faults that are inclined to the top surface of the sample, the reciprocal lattice point gets spiked in a crystallographic direction that is perpendicular to the plate, stacking fault or twin interface. The diffraction spots show streaks of intensity running from one diffraction spot to the other corresponding to the particular crystallographic direction. The streaks will be of shorter length in the case of plates or disc-shaped precipitates.

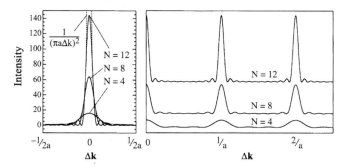

Fig. 5.15 Crystal size effect on diffracted intensity [after Brent Fultz and James Howe, Transmission Electron Microscopy and Diffractometry of Materials, 2013, Springer-Verlag Berlin Heidelberg. With permission]

Table 5.1 Shape and size effects of a crystal on diffracted intensity

No of Dimensions in which the crystal is small	Shape of the crystal	Relrod extension	Diffraction effects
Three	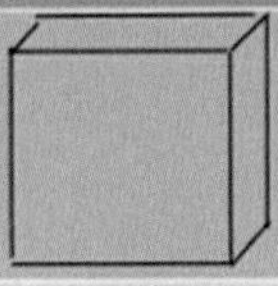	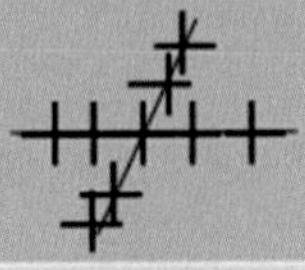	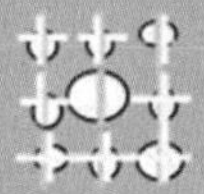
Three			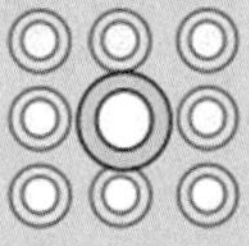
One		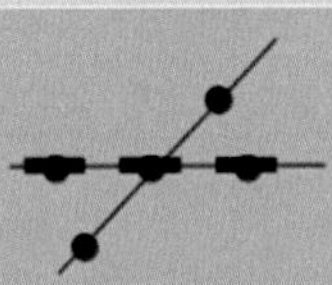	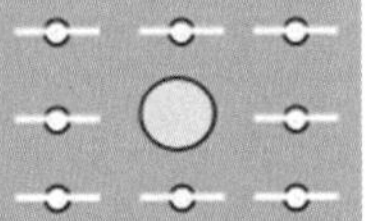
Two		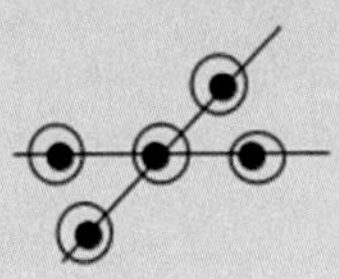	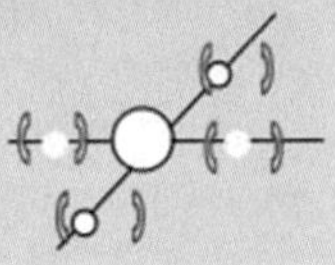

When rod- or needle-shaped precipitates are considered, the shape factor would lead to concentric discs and rings in the reciprocal space around the long axis of the needle/rod. The diffraction spots would show arcs or sections of circles depending on the angle of cut made by the Ewald sphere.

Structure Factor:

Let us consider Eq. (5.12) once again to assess the contribution of structure factor $f(\theta)$ to the diffracted intensity for Bravais lattices in 3-D crystals. We have seen that each atom of the unit cell (or a group of atoms associated with each equipoint of the unit cell) scatters incident electron in all directions with a scattering strength f, which is defined as atomic scattering factor. The phase value of the scattered wave is determined by the position of the particular atom in the unit cell. The algebraic sum of scattered waves from all the atoms of a unit cell with reference to any arbitrary

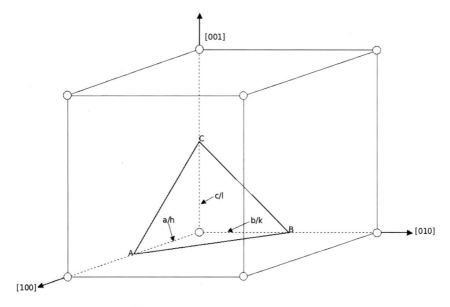

Fig. 5.16 Structure Factor $\overline{F}_{hkl}$ for a primitive cell

origin will yield the resulting amplitude, i.e. the structure factor $\overline{F}$. For a chosen crystallographic plane it will be $\overline{F}_{hkl}$. Since intensity of the diffracted beam is the only observable quantity on the screen, we are interested in computing the square of the structure factor $|\overline{F}_{hkl}|^2$, which gives the intensity. In a typical unit cell as depicted in Fig. 5.16 there is a scatterer at 1,0,0 coordinate position along the [100] direction.

The scattered wave from it will be exactly one wavelength (or integer multiples thereof) out of phase with that at the origin when it is first-order reflection (with a phase value 2π). The phase value will be $2\pi h$ if the scattered beam is h-order reflection. Similarly in the other two perpendicular directions [010] and [001] with phase values $2\pi k$ and $2\pi l$, respectively.

i.e. $\phi = 2\pi$ (hu+kv+lw) where u, v, w are the fractional coordinates of the scatterers. Thus, if there are N scatterers in the unit cell their phase-amplitude diagram can be constructed as in Fig. 5.17.

The structure factor $\overline{F}$, then, is the sum of scattered wave amplitudes $f_1, f_2, ... f_n$ and phase angles $\phi_1, \phi_2, ... \phi_n$.

$$|\overline{F}_{hkl}|^2 = \left[\sum_n f_n Cos\phi_n\right]^2 + \left[\sum_n f_n Sin\phi_n\right]^2$$

$$= \left[\sum_n f_n Cos 2\pi(hu_n + kv_n + lw_n)\right]^2 + \left[\sum_n f_n Sin 2\pi(hu_n + kv_n + lw_n)\right]^2$$

$$(5.19)$$

Fig. 5.17 Phase-amplitude
diagram

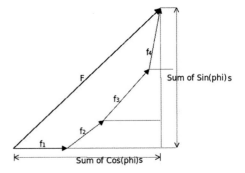

For centrosymmetric crystals (crystal possessing centre of inversion or in which for
every atom at x, y, z there is an atom at −x, −y, −z in the unit cell) the Sine term
in the Equation vanishes (as $Sin(-\theta) = -Sin(\theta)$). The expression is also universally
applicable for all crystals. When the phase factors are expressed as complex numbers
it takes the form:

$$\overline{F} = \left[\sum f_n exp[2\pi i (hu_n + kv_n + lw_n)]\right] \qquad (5.20)$$

and

$$|\overline{F}|^2 = \sum f_n^2 [Cos2\pi (hu_n + kv_n + lw_n) + Sin2\pi (hu_n + kv_n + lw_n)]^2.$$

Let us apply this to the case of a simple cubic structure which has one atom per
unit cell.

$$\overline{F}_{scc} = \sum f_n exp[2\pi i (hu_n + kv_n + lw_n)],$$

where u, v, w take the values 0, 0, 0 and

$$|\overline{F}|^2 = f_n^2 [Cos^2 2\pi (0) + Sin^2 2\pi (0)]$$
$$= f_1^2 [1]$$

Therefore, all (hkl) reflections are permitted.

A second example is that of f.c.c. unit cell in which there are four atoms, one
at cube corners with (0,0,0) coordinates and three others at $(\frac{1}{2}, \frac{1}{2}, 0)$, $(\frac{1}{2}, 0, \frac{1}{2})$ and
$(0, \frac{1}{2}, \frac{1}{2})$ face centres (see Fig. 5.18).

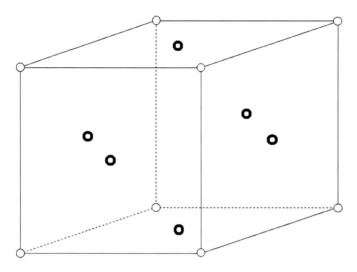

Fig. 5.18 Face centred cubic cell

$$\overline{F}_{fcc} = \sum f exp[2\pi i(0)] + f exp\left[2\pi i\left(\frac{h}{2} + \frac{k}{2} + 0\right)\right] + f exp\left[2\pi i\left(\frac{h}{2} + 0 + \frac{l}{2}\right)\right]$$
$$+ f exp\left[2\pi i(0 + \frac{k}{2} + \frac{l}{2})\right]$$
$$= f(1 + exp[\pi i(h + k)] + exp[\pi i(h + l)] + exp[\pi i(k + l)])$$

Indices h, k, l may take odd or even values. Since they are in pairs, their sum will be an even number for unmixed integers. Therefore, the exponential terms return integral phase differences of $2n\pi$ and

$$\overline{F}_{fcc} = f[1 + 3] = 4f \text{ and}$$
$$|\overline{F}|^2 = 16f^2$$

For mixed indices, say h and k are even and l is odd,

$$\overline{F}_{fcc} = f[1 + 1 - 1 - 1] = 0 \text{ and } |\overline{F}|^2 = 0.$$

Therefore, reflections with either even or odd h, k, l values are only permitted for face centred cubic lattice.

Note that: (1) though we used the example of a face centred cubic lattice, any parameter related to the symmetry of the lattice has not been included in the expression for $\overline{F}$. Thus structure factor $\overline{F}$ is independent of the shape and size of the unit cell, unlike the **Shape Factor**. Its value depends only on the location of the scatterers within the unit cell. The scatterers may not be single atoms but a group in some crystals.

(2) although all reflections are permitted in the case of a simple cubic lattice or reflections with unmixed indices in f.c.c., the intensity of all the permitted reflections may not be equal, particularly in X-ray diffraction. The intensity of individual reflections depends on several factors such as multiplicity etc. For details you may consult any standard book on X-ray diffraction.

5.5   Indexing of Electron Diffraction Patterns

A typical diffraction pattern, from a specimen that is randomly oriented with respect to the incident beam, is shown in Fig. 5.19.

The diffraction spots in this pattern bear no specific relation to each other. In such cases, the spots can be indexed if the crystal structure of the material is known, using the camera constant (CC) of the microscope from the relation

$$CC = rd$$

where r is the distance from the transmitted beam to the selected spot and d is the interplanar spacing to be determined for the selected diffraction spot. The above relation can easily be derived from the reciprocal space representation of Bragg's law (see Fig. 5.12) in which two similar triangles Ok_oK' and OAB can be identified. According to the principle of similar triangles

$$\frac{Ok_oK'}{k_ok'} = \frac{OA}{OB}$$

$$\frac{\frac{1}{\lambda}}{\frac{1}{d}} = \frac{L}{r}$$

$$i.e.\ \lambda L = rd \tag{5.21}$$

Fig. 5.19 Typical diffraction pattern with a randomly oriented incident beam

Fig. 5.20 Typical diffraction
patterns along three different
zone axes of a cubic crystal
(Schematic)

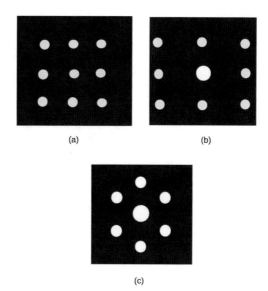

(a)   (b)

(c)

where λ is the wavelength of the electrons used for diffraction, L is the Camera
length. λL remains constant for a chosen accelerating voltage in the microscope and
chosen magnification of the diffraction lens.

When the electron beam is oriented parallel to any desired crystallographic direction by the manipulation of the specimen stage in the microscope, a highly symmetric
diffraction pattern emerges around the transmitted spot. All the crystal planes that
are parallel to the selected crystallographic direction diffract to give rise to this symmetric pattern, Fig. 5.20 shows three such patterns from an unknown crystal.

The three patterns a,b and c exhibit 4-fold, 2-fold and 6(3)-fold symmetry, respectively. When we examine the 14 Bravais lattices in Crystallography Tables, we notice
that 4-fold axis of symmetry is present in cubic and tetragonal lattices only. Let us
assume that the lattice under investigation is cubic as first choice, since many commonly used metals and alloys have cubic structure. The pattern represented in (b)
shows 2-fold axis of symmetry which is present in cubic, tetragonal, orthorhombic
as well as hexagonal lattices/hexagonal close packed structures. The pattern in (c)
displays a 6-fold axis of symmetry or a 3-fold axis of symmetry that appears as
a 6-fold due to the operative Friedel's law in electron diffraction. While they are
indistinguishable in parallel beam electron diffraction, they can be delineated by
convergent beam electron diffraction technique which is explained in a later section
of the chapter. Therefore, from patterns in (a) and (c) we can reduce our choices of
the lattice to cubic. Let us now take help of stereographic projection of cubic crystals
(see Sect. 5.2.2 and Appendix A). The standard projection of a cubic crystal along
[001] shows 4-fold symmetry and hence the poles projected on the equatorial circle
are the reflections observed in the diffraction pattern at the respective angles along
[001] zone axis unless until they are forbidden by the structure factor considerations

Fig. 5.21 Typical diffraction patterns along three different zone axes of a cubic crystal (Schematic)

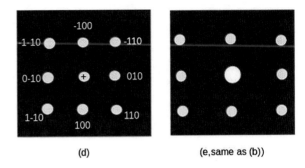

(d)   (e, same as (b))

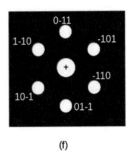

(f)

(in which case they may be even multiples of the indices associated with each of the poles.). As per this procedure, the pattern is indexed tentatively as shown in Fig. 5.21.

The 6-fold symmetry pattern can also be tentatively indexed as represented in Fig. 5.21f. There is no unique choice available from the standard cubic projections for the 2-f pattern in Fig. 5.21b. Besides this, indexing of the other two patterns is also tentative. Therefore, for uniquely identifying the Bravais lattice corresponding to the three diffraction patterns we adopt a method called ratio method at this stage of analysis.

5.5.1 Ratio Method

Let us identify three non-coplanar vectors that span the first three nearest neighbour spots from the transmitted spot in the pattern, (Fig. 5.23a). Ratio of their magnitudes measured in mm are obtained as follows:

$$\frac{r_2}{r_1} = \frac{n}{m} \tag{5.22}$$

$$\frac{r_3}{r_1} = \frac{o}{m} \tag{5.23}$$

Since we have assumed the lattice to be cubic we will examine the quadratic relations between d values and hkl for all the permitted Bravais lattices in cubic system.

$$a = d_{hkl}(\sqrt{(h^2 + k^2 + l^2)}) \qquad (5.24)$$

$$\text{i.e.} \frac{1}{a^2} = \frac{1}{d_{hkl}^2(h^2 + k^2 + l^2)}$$

$$\text{or } d_{(hkl)}^2 = \frac{a^2}{\sum h^2} \qquad (5.25)$$

We also know from Sect. 5.2 that structure factor decides which of the planes diffract electrons. These conditions are listed in Sect. 5.4.

We have derived the equivalent of Bragg's law in reciprocal space as

$$CC = rd_{(hkl)}$$

$$CC = r_1 d_1 = r_1 \frac{a}{\sqrt{\sum h_1^2}} = r_2 \frac{a}{\sqrt{\sum h_2^2}}$$

$$\frac{r_2}{r_1} = \frac{\sqrt{(h_1^2 + k_1^2 + l_1^2)}}{\sqrt{(h_2^2 + k_2^2 + l_2^2)}} \qquad (5.26)$$

In the case of our example pattern in Fig. 5.21(d) m = n. Therefore, $\frac{r_2}{r_1} = 1$. We can index the pattern using Eqs. (5.22), (5.2) and (5.26) and Table 5.2 in a precise way as shown in Fig. 5.22g. Mutual compatibility of the assigned indices needs to be verified by vectorial additions as well as the angular relations need to be satisfied experimentally and by calculation using Cosine formulae given in Appendix C.1. Finally the zone axis is obtained by taking cross product of any pair of non-collinear vectors, say h_1, k_1, l_1 and h_2, k_2, l_2 and represent the same as in Fig. 5.22g. During the above analysis it would be realised that diffraction patterns (d) and (f) are not unique to any one cubic structure and can be correctly indexed on the basis of either b.c.c. or f.c.c. structure. Therefore, it is the pattern given in (e) (i.e. pattern (b)) that uniquely decides which of the two is correct. Keeping this in view, the remaining two patterns are indexed in a similar way and are shown in Fig. 5.22h and i.

The lattice can be uniquely identified as b.c.c. Ratio method is the most general method and is applicable to all the 14 Bravais lattices. The cases of low-symmetry lattices such as hexagonal are, however, more complicated due to differing c/a ratio from crystal to crystal or the non-orthogonality of basis vectors in monoclinic and triclinic crystals.

Polycrystalline samples: Diffraction patterns from polycrystalline samples show either continuous or spotty concentric rings surrounding the transmitted spot in a

Table 5.2 Reflections permitted by structure factor

hkl	$\sum h^2$	Bravais lattice		
		s.c.	b.c.c.	f.c.c.
100	1	1		
110	2	2	2	
111	3	3		3
200	4	4	4	4
210	5	5		
211	6	6	6	
	7			
220	8	8	8	8
221,300	9	9		
310	10	10	10	
311	11	11		11
222	12	12	12	12
320	13	13		
321	14	14	14	
	15			
400	16	16	16	16

systematic way (Table 5.3). Since the incident electron beam samples a small volume of the thin specimen, the size of the crystallites should be such that at least a thousand grains are present in the illuminated volume. In order to index the pattern given in Fig. 5.23b, we measure once again the lengths $r_1, r_2, ...r_n$ and tabulate the ratios $\frac{r_n}{r_1}$.

The experimentally obtained ratios are compared with those calculated from Table 5.2, which ever series matches well with the experimental one within permitted errors is taken as the Bravais lattice of the poly crystalline sample studied. In the present example it is an f.c.c. lattice. The match is not very good since it is a schematically drawn pattern but not an actual diffraction pattern from a polycrystalline specimen. Note that the sequence of ratios is same for simple cubic lattice and body centred cubic lattice till the sixth reflection. Hence in order to distinguish between the two, one must observe and measure up to seventh ring in the pattern.

Camera constant method: If the crystal structure of the material being investigated is already known and we want to index the corresponding diffraction pattern for any other application such as determining the Burgers vector of dislocations etc. we need not adopt such a elaborate procedure as the ratio method, but use Eq. 5.21. To calculate the d values corresponding to the diffraction spots observed in the pattern. Indices are then assigned d values by matching them with those given in (JCPDS) Powder Diffraction Data files, mutual compatibility verified and zone axis determined.

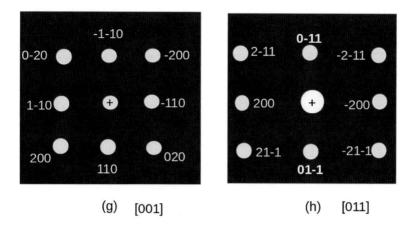

(g)   [001]

(h)   [011]

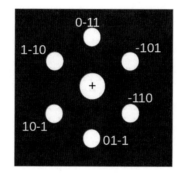

(i)   [111]

Fig. 5.22 Completely indexed diffraction patterns along three different zone axes of a cubic crystal

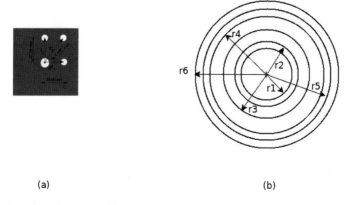

(a)

(b)

Fig. 5.23 Indexing of polycrystalline patterns

Table 5.3 Ratios of permitted reflections in f.c.c. crystals

Sl. no.	$\frac{r_n}{r_1}$ (exp.)	$\frac{r_n}{r_1}$ (calc.)	hkl
1			111
2	1.33	1.15	200
3	1.66	1.63	220
4	2.22	1.91	311
5	2.5	2.0	222
6	2.77	2.31	400

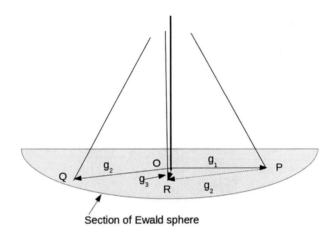

Section of Ewald sphere

Fig. 5.24 Origin of double diffraction spots

Double Diffraction: It is a phenomenon in which a strong beam of electrons that is diffracted by a set of $h_1k_1l_1$ planes at Bragg angle θ_1 acts as a primary beam for another set of planes $h_2k_2l_2$ located differently at θ_2 (albeit belonging to the same zone axis), a new diffraction vector becomes operative leading to a third beam at θ_3 which otherwise is forbidden by structure factor considerations. This phenomenon best be illustrated as in Fig. 5.24. Points O, P, Q and R are coplanar reciprocal lattice points corresponding to the crystal under consideration and the shaded region is the section of Ewald sphere in that orientation.

Here, O represents the intersection of the incident vector $\mathbf{k}_o$ with the reciprocal lattice and P represents another reciprocal lattice point corresponding to (hkl) plane through which as well, the Ewald sphere passes. Therefore, a diffracted beam will be generated which may act as a primary beam and excite another diffracted beam at R. Here, r is at a vectorial distance of $|\mathbf{g}_3|$ from P. Since the Ewald sphere is also intersecting the reciprocal lattice section at Q corresponding to $|\mathbf{g}_2|$ vector representing $(h_2k_2l_2)$, Q can also act as primary beam to generate a diffracted beam at R, which is at a vectorial distance of $|\mathbf{g}_1|$ from Q. In other words, diffracted intensity is reinforced at R, i.e. $(h_3k_3l_3)$ plane, which is otherwise excluded by the structure factor (structure factor $F = 0$). Therefore a weak diffracted intensity will be observed from the plane $(h_3k_3l_3)$ provided

$$h_3 = h_1 \pm h_2$$
$$k_3 = k_1 \pm k_2$$
$$l_3 = l_1 \pm l_2$$

Now consider the case of a face centred cubic lattice structure, the structure factor calculations put the conditions that h, k, l should be unmixed indices for diffracted intensity to be observed. Therefore, the above equation also should satisfy the same condition, i.e.

$$\text{if } h_1 = 2, k_1 = 0 \text{ and } l_1 = 0$$
$$h_2 = 0, k_2 = 2 \text{ and } l_2 = 0$$

then

$$h_1 + h_2 = 2,$$
$$k_1 + k_2 = 2, \text{ and}$$
$$l_1 + l_2 = 0.$$

Therefore, the resulting double diffraction spot is also a permitted reflection (not excluded by **F**) and by no combination of **g** vectors can we reach (010) or (110) reflections, marked by red circles. Let us consider a diamond crystal which has a cubic lattice with a different structure. This structure also follows the same condition of unmixed h, k, l together with an additional condition that h, k, l be unmixed and h+k+l≠ 4n+2 for diffraction to occur. Taking $(h_1k_1l_1)$ as $1\bar{1}1$ and $(h_2k_2l_2)$ as $\bar{1}11$ we can demonstrate that there will be weak intensity present in the (002) reflection which is otherwise forbidden according to structure factor. A similar occurrence of double diffraction in the case of h.c.p. structure can also be worked out (Fig. 5.25).

The observed difference between f.c.c. and other structures can be understood by modelling double diffraction under dynamical conditions (De Graef 2003). The inference is that double diffraction does not occur in case of crystals that exhibit

Fig. 5.25 No Double diffraction in case of f.c.c

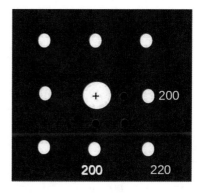

systematic absences due to centre of symmetry. For others it should be worked out by reference to the space group of the crystal under consideration.

Diffraction patterns from quasicrystalline materials such as that shown in Fig. 5.26a cannot be indexed using the 'Ratio Method' due to the quasiperiodicity of the spacing of diffraction spots in such patterns, e.g. the row of successive spots indicated by arrows. When we measure the distances from the transmitted beam, we will find that each of the successive spacings is not an integer multiple of the first spacing from the transmitted beam. In the corresponding real space, the scatterers are at distances $a + b\tau$, where a and b are integers and τ is an irrational number, $\dfrac{1 + \sqrt{5}}{2}$. The symmetry present in the diffraction pattern can be inferred though. In the present case we may interpret the diffraction symmetry to be 10-fold, but actually it is 5-fold symmetry appearing as 10-fold due to Freidel's law, though both of them are crystallographically forbidden rotational symmetries in periodic crystals. Hence we do not have any standard stereograms to help us make any guess about the reciprocal space. Sastry et al. (1978) observed similar patterns in Al-Pd and $Al_{60}Mn_{11}Ni_4$, and interpreted them as arising from a new ordered phase that exhibits pseudopentagonal symmetry (Fig. 5.26b) on the basis of high resolution images of the phase (Fig. 5.26c). Thus it is all the more important to explore the reciprocal space by systematic tilts of the specimen in the goniometer of the TEM. This was done by Shechtman et al. (1984) and discovered the quasicrystalline phase possessing the 5-fold, 3-fold and 2-fold symmetries and rightly mapped the reciprocal space. This has broken the dictum existing thus far that rotational symmetries such as 5-fold, 10-fold etc. cannot be compatible with periodicity of crystalline materials. Indexing of such patterns is still not simple as all the three diffraction patterns (5-f, 3-f and 2f diffraction patterns) exhibit quasiperiodicity (i.e. though 3-f and 2-f are crystallographically permitted). Structure of the quasicrystalline phase and consequently, indexing of the diffraction patterns from them were interpreted in 6-D space in which the Miller indices would have six integers (Cahn et al. 1986). A unified indexing scheme was developed by Mandal et al. (2004) to index diffraction patterns from 3-D, 2-D and 1-D quasicrystals. A detailed discussion of these schemes is beyond the scope of the book. They require a mathematically more complex procedure. Interested readers are referred to the publications by Mandal and Lele (1989) and Lu et al. (2002).

5.5.2 Kikuchi Lines and Kikuchi Patterns

When a high energy beam of electrons is incident on a slab of crystal, majority of the electrons are transmitted and the remaining are elastically scattered giving rise to diffracted beams at Bragg angles. In case the crystal slab is thick some of the incident electrons suffer inelastic scattering as at 'O' in Fig. 5.27 and form a diffuse background intensity in the diffraction pattern on the screen. Since these electrons would have lost only a small part of their kinetic energy, they are likely to be Bragg diffracted by the same crystal planes which diffract the elastically scattered electrons.

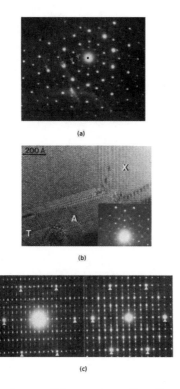

Fig. 5.26 Quasicrystalline patterns **a** 5-f symmetry pattern from Al-Mg-Zn alloy [after P. Ramachandrarao and G. V. S. Sastry, Pramana 25 (2), (1985) L225–L230. With permission], **b** Pattern showing 5-f symmetry in Al-Pd alloy along with a HREM image [after G. V. S. Sastry, C Suryanarayana, M. Van Sande, G. Van Tendeloo, Materials Research Bulletin, 13 (10), (1978) 1065–1070.With permission], **c** Pattern from the Al-Pd alloy as in **b** showing periodicity along one direction (horizontal in the left Figure and that from an $Al_{60}Mn_{11}Ni_4$ alloy [after G. V. S. Sastry, C. Suryanarayana, M. Van Sande, G. Van Tendeloo, Materials Research Bulletin, 13 (10), (1978) 1065–1070. With permission]

However, they (the inelastically scattered electrons) differ in their wavelength from the elastic ones which may be ignored considering the very small Bragg angles in electron diffraction (only in milliradians). Consider two beams OA and OB arising out of the inelastic event that occurred at 'O'. Since OA is satisfying Bragg angle, it is diffracted elastically in the direction A'. The beam OB is also incident at the same Bragg angle but on the plane m'n' on the opposite side and is diffracted to B' in the same direction of OA. To begin with, the beam OA is having higher energy in comparison to OB, but the net energy diffracted in the direction OA(B') is smaller than that in the direction of OB(A'). Hence, on an observation plane (such as microscope screen) the beam B' will appear less intense than the background and the beam A' will appear brighter than the background. Hence, A' is termed as excess Kikuchi line and B' as deficit Kikuchi line, named after S.Kikuchi, who discovered them

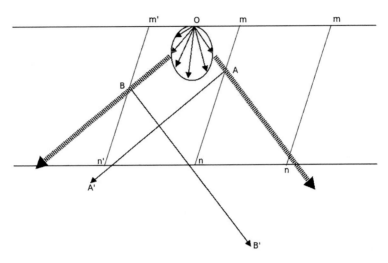

Fig. 5.27 Origin of Kikuchi lines

in electron diffraction in the year 1928. Kikuchi lines are observed in diffraction
patterns obtained from either thick specimens or thick regions of thin specimens
where elastically scattered electrons begin to loose their intensity and give rise to
inelastic scattering as shown in Fig. 5.27. In extreme cases of thickness, the entire
pattern is formed by the Kikuchi lines only.

In reality these are not straight lines as they appear to be, but are segments of
hyperbolae (intersection of the screen) and appear as pairs on either side of a diffraction spot. In fact, they 'stick' to one **g**-vector (diffracting plane) and move about the
corresponding diffraction spot that occurred due to initial elastic scattering. it can be
easily deduced from the geometry of the event presented in Fig. 5.27, that is when
they are symmetrically inclined, the bright or excess line passes through the centre
of the corresponding **g** spot and the dark or deficit line passes through the centre of
the transmitted spot. These features can be seen in the diffraction patterns illustrated
in the following Fig. 5.28.

When the complete zone axis pattern is symmetrically excited, we observe a
symmetric map as given in Fig. 5.29. When only a map is formed without spots it
appears similar to a bend centre in real-space image, either BF or DF. Maps help in
identifying the true symmetry of the lattice as is the triad axis in the diamond lattice
of Si. An entire Kikuchi map of the reciprocal lattice of the crystal being investigated
is displayed when the diffraction lens is operated at a lower magnification. These
maps also can be indexed just as the diffraction patterns can be. A typical indexed
map is shown in Fig. 5.30 (Fultz and Howe 2013).

It aids in reorienting a specimen along a desired zone axis—often required for
defect characterisation, etc. Kikuchi maps are similar to the Electron Back Scattered
Diffraction (EBSD) patterns except that the later are formed by elastically scattered
electrons.

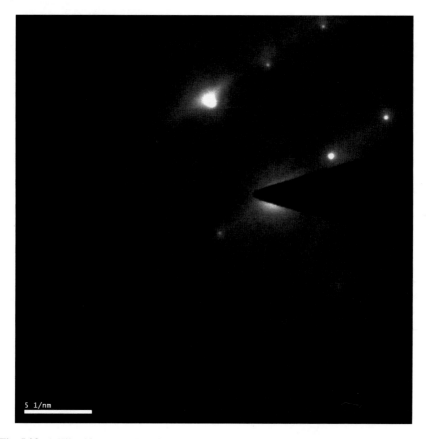

Fig. 5.28 A Kikuchi pattern. Note that it is almost a map with very weak spots

Fig. 5.29 A Kikuchi map of
[111] zone axis of Silicon
[after G. Thomas, M. J.
Goringe, Transmission
Electron Microscopy of
Materials
(Wiley-Interscience, New
York, 1979). With
permission]

Fig. 5.30 Indexed Kikuchi
map [after Brent Fultz and
James Howe, Transmission
Electron Microscopy and
Diffractometry of Materials,
2013, Springer-Verlag,
Berlin, Heidelberg. With
permission]

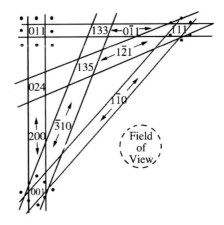

A detailed description of the Kikuchi lines on the basis of dynamical theory of electron diffraction is given by many researchers in the context of EBSD maps. Kainuma (1955). A note of caution while reading the earlier literature is that the excess lines are described to be appearing dark by earlier authors, because these left darker imprint on the photographic film. The reader may also find it difficult to grasp why the Bragg diffracted spots behave differently from the Kikuchi lines, when the latter are also formed by Bragg diffraction of the inelastically scattered electrons.

5.6 Determination of Orientation Relationships

5.6.1 Matrix-Twin Orientation Relation

A crystal lattice can take different orientations within a grain in a polycrystalline specimen due to either twinning or ordering. We will discuss the case of ordering in the next section. There are different ways in which twins can be classified and we adopt the following classification.

1. Growth twins. 2. Twins formed during recrystallization process of deformed specimens. 3. Transformation twins.

Growth twins: Twinning of the crystal takes place during growth of many minerals. Portions of crystal grow on an initially nucleated one by taking different growth variants that are twin related planes. The aggregate is a bulk crystal finally and is not amenable to characterisation as such due to its macroscopic size. We will demonstrate the same effect in case of irrational twins though. These twins as shown in Fig. 5.31 result in a rapidly solidified Al-Cu-Fe-Ni alloy as the nodule tries to grow in the form of a cluster of plates by a process of twinning. The twin plane in this case is an irrational plane.

Fig. 5.31 Irrational twins in rapidly solidified Al-Cu-Fe-Ni alloy, [after R. K. Mandal, G. V. S. Sastry, S. Lele, S. Ranganathan, Scripta metallurgica et materialia 25(6), (1991) 1477–1482. With permission]

Recrystallization twins: Two portions of a lattice within a grain are considered as twinned if the lattice points of a region and hence the atoms located at those points, happen to be a mirror reflection of the other. The matching plane across which this mirroring takes place is known as the twin plane. Such types of twins are known as reflection twins or mirror twins. Twinning can also take place when portion of the lattice is rotated about an axis, which is either parallel or perpendicular to the plane by a 2-fold, 3-fold or 4-fold operation. Such twins are known as rotational twins. Twinning operation can be an inversion in some crystals when such inversion centre is not present already as a symmetry element of the lattice. In cubic crystals like f.c.c. and also h.c.p.the above conditions are already met as part of the symmetry possessed by the lattice and as such are called compound twins. An example of mirror twinning in f.c.c. is given in Fig. 5.32.

The close packed (111) plane is the twinning plane and $[1\bar{1}0]$ is the twinning axis. The mirror twins often have coherent twin interfaces which are atomistically thin. A diffraction pattern is obtained in the case of f.c.c. mirror twin from regions that include both matrix and twin. It is evident that the pattern is a composite of the two variants which are in $[1\bar{1}0]$ orientation.

Deformation twins are formed in f.c.c alloys on the same plane, i.e. on 111 planes. Since two equivalent positions are possible for these close packed planes, the deformation twins cannot grow much larger in size under continued deformation and remain thin. The diffraction effects are similar to those of mirror twins.

Transformation twins: It is generally known in crystallography that when a high-symmetry crystal phase at high temperature undergoes a polymorphic transformation to a low temperature phase, the lost symmetry elements manifest themselves as transformation twins in the low temperature phase. The $YBa_2Cu_3O_{7-\delta}$ high T_c superconductor exhibits twinning on a-b plane (the twin plane being (110)) upon cooling from the tetragonal triple perovskite phase that exists at high temperature. Note the distinction between a and b lattice parameters is no longer there in a tetragonal crystal lattice. Nearly equally-wide twins can be noticed in the electron micrograph as alternate dark and bright bands in Fig. 5.33. The corresponding diffraction pattern (in (b)) shows splitting of the 110 spots along the diagonal with increasing linear separation as you move out from the transmitted spot, a manifestation of the twins.

Fig. 5.32 **a** Schematic
representation of twins in
f.c.c. crystal **b** Diffraction
pattern in [01$\bar{1}$] zone axis
and **c** Twins in Ni-Fe-based
superalloy [after A.
Bhattacharyya, G. V. S.
Sastry, V. V. Kutumbarao,
Journal of materials science
34(3), (1999) 587–591. With
permission]

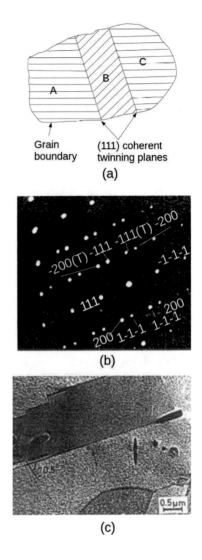

(a)

(b)

(c)

5.6.2 *Precipitate-Matrix Relation*

In a multi-phase system, a precipitate may bear a specific orientation relation with
the matrix as the latter offers a matching plane where the atomic coordination is
nearly the same in both precipitate and matrix. When more than two precipitates
are present, usually one of them bears a relation with the matrix although there are
exceptions to this (γ' and $M_{23}C_6$ carbide in γ phase of a nickel-base superalloy). A
diffraction pattern taken from an IN617 alloy showing this relationship is presented
in Fig. 5.34.

Fig. 5.33 Twins in YBaCu oxide and the corresponding diffraction pattern in [110] zone axis [after G. V. S. Sastry, R. Wödenweber, H. C. Freyhardt, Journal of Applied Physics 65(10), (1989) 3975–3979. With permission]

The orientation relationship and the possible no. of variants can be established by the superposition of the respective stereograms. In the present case of IN617 alloy, it is a cube-on-cube relation of the carbide with the matrix. That is

$[001]M_{23}C_6 \parallel [001]$ γ-iron
$(100)M_{23}C_6 \parallel (100)$ γ-iron
$(010)M_{23}C_6 \parallel (010)$ γ-iron

For establishing a precise orientation relationship, reciprocal space of the matrix phase needs to be carefully explored beginning from high-symmetry zone axes to the commonly observed low-symmetry axis. The microstructure also helps in exploring any possible existence of O-R by showing sharp planar interfaces between the matrix and precipitate.

Other well-known orientation relationships are Kurdjumov-Sachs relation and Nishiyama-Wasserman relation in the case of martensite (ϵ martensite, c.p.h. structure) and phases with f.c.c. and b.c.c. structures in steels. The diffraction patterns exhibiting these relationships are schematically shown in Fig. 5.34b and c. The precise determination of these relations is best done taking help of the stereographic projections (Refer to Andrews et al. 1971).

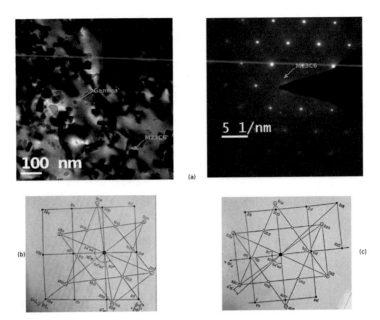

Fig. 5.34 **a** Cube-on-cube orientation relation of $M_{23}C_6$ carbide in IN617 nickel-base superalloy, BF image and corresponding Diffraction Pattern in [110 orientation] [after Ch. Visweswara Rao, Ph.D. Thesis, IIT (BHU), Varanasi, India, 2020. With permission]. **b** K-S Orientation relation [after K. W. Andrews, D. J. Dyson, and S. R. Keown, Interpretation of Electron Diffraction Patterns, Adam Hilger Ltd., London, 1971. With permission] and **c** N-W Orientation relation [after K. W. Andrews, D. J. Dyson, and S. R. Keown, Interpretation of Electron Diffraction Patterns, Adam Hilger Ltd., London, 1971. With permission]

5.7 Effect of Ordering of the Crystal

A binary solid solution of B in A, if ideal, will have no distinction of sites occupied by the B atoms in the lattice of A and vice versa. Usually this is the structure prevailing at high temperature and upon cooling the solid solution below a temperature T_c, the B atoms occupy specific sites such as body centre or face centre, if the chemical activities of the components show negative deviation from ideal behaviour. The substitutional solid solution is said to be ordered. Interstitial solid solutions may also get ordered by the occupation of specific interstitial sites by the B atoms.

Ordered structures based on b.c.c.: The body centred cubic lattice has two lattice points $0,0,0$ and $\frac{1}{2},\frac{1}{2},\frac{1}{2}$. Therefore binary solid solutions having a b.c.c. lattice can undergo ordering with the B atoms either at $\frac{1}{2},\frac{1}{2},\frac{1}{2}$ or $0,0,0$. The structure factor equation (Eq. 5.19) has now two distinct terms, one for A atoms and the other for B atoms.

i.e. $(\overline{F}_{hkl})^2 = [f_A Cos2\pi(h0 + k0 + l0) + Cos2\pi(h\frac{1}{2} + k\frac{1}{2} + l\frac{1}{2})]^2 + [f_A$ $Sin2\pi(h0 + k0 + l0) + f_B Sin2\pi(h\frac{1}{2} + k\frac{1}{2} + l\frac{1}{2})]^2$, assuming that all B atoms are at $\frac{1}{2}, \frac{1}{2}, \frac{1}{2}$ positions.

$$i.e. (\overline{F}_{hkl})^2 = [f_A + f_B Cos\pi(h+k+l)]^2 + [f_B Sin\pi(h+k+l)]^2.$$

When h+k+l = even, only the first [ ] survives and
$F^2 = [f_A + f_B]^2$.

When h+k+l is odd

$$F^2 = [f_A - f_B]^2.$$

Therefore, when (h+k+l) is even, the reflections are similar to b.c.c. and are permitted reflections of b.c.c.; and when (h+k+l) is odd the intensity is non-zero but weak.

An isotype showing this type of ordering by a b.c.c. structure is CsCl where chlorine atoms get ordered on $\frac{1}{2}, \frac{1}{2}, \frac{1}{2}$ positions. The structure somewhat resembles b.c.c. but the lattice is not body centred. A typical diffraction pattern in [100] zone axis appears similar to that shown in Fig. 5.35.

The example shown here is that of a AlCu phase metastably occurring in rapidly solidified Al-Cu alloy. The intense reflections are common to the disordered and ordered states of the phase, while the weak (010) reflection arises due to ordering. As is clear from indexing, the lattice is no longer body centred cubic and intensity enhancement of the (020) reflection is due to the additive contribution of the atomic scattering factors of Al and Cu. Further, the reflection due to ordering will be visible only if the two atomic scattering factors differ appreciably. The above treatment is true for any body centred lattice and it need not be cubic. Isotypes are β'CuZn, β AlNi, etc.

Ordered structures based on f.c.c.: Face centred cubic structure can undergo mainly two types of ordering, one where the cube corners are occupied by one type of atoms (0, 0, 0 position) and the face centres $(0, \frac{1}{2}, \frac{1}{2})$, $(\frac{1}{2}, \frac{1}{2}, 0)$, $(\frac{1}{2}, 0, \frac{1}{2})$ by the other (Cu$_3$Au,L1$_2$ type) and the other where cube corners and two opposite face centres by one type of atom and the remaining face centres by the other (CuAuI, L1$_0$). This type of ordering results in a tetragonal structure for the ordered state. Using the Eq. 5.19 we can arrive at the permitted reflections for the fundamental and superlattice reflections.

Fig. 5.35 Diffraction pattern
from ordered b.c.c phase

L1₂ structure: Cube corners are Au atoms and contribute to the intensity of fundamental reflections as

$$F = 3f_{Cu} + f_{Au} \text{ for h, k, l odd or even.}$$

The face centres are exclusively occupied by Cu atoms and contribute to intensity of superlattice reflections as

$$F = f_{Cu} - f_{Au} \text{ for mixed h, k, l.}$$

The diffraction pattern in the commonly observed orientation is given in Fig. 5.36.
L1₀ structure: The ordered structure is a base centred tetragonal with Cu atoms at cube corners and base centres while Au atoms are at the side centres. Contribution to the intensity of fundamental reflections is as
$F = 2f_{Au} + f_{Cu}$ for mixed h, k, l.
Contribution to the intensity of superlattice reflections is as
$F = 2f_{Au} - f_{Cu}$ for h, k, l such that l odd and others even or l even and others odd.
The diffraction patterns corresponding to this type of ordering are shown in Fig. 5.37. The isotypes for L1₂ are α'-AlCo₃, FeNi₃, etc. and for L1₀ are AlTi, FePt, etc.
Short Range Order (SRO): When ordering is incomplete either due to temperature or due to non-stoichiometric composition of the alloy, short range order (SRO) prevails in the alloy. A simplest model due to Bragg and Williams defines SRO parameter S, which at the critical temperature and composition reaches a value 1 indicating complete order, and reaches a zero value above T_c indicating complete disorder for all compositions. When S has fractional value diffuse intensity is observed in the

Fig. 5.36 L1₂ pattern Cu₃Au Type, Ni₃(Ti,Al) [111] Zone: [after Ch. Visweswara Rao, N. C. Santhi Srinivas, G. V. S. Sastry, Vakil Singh, Mater. Sci. Eng. A, 765 (2019) 138286. With permission]

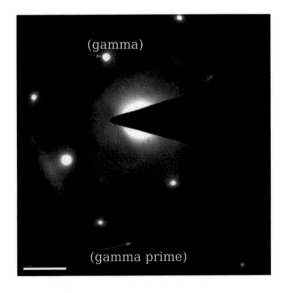

(gamma)

(gamma prime)

Fig. 5.37 L1$_0$ pattern
CuAuI type in Ni$_4$Mo [after
S. Banerjee, K. Urban and
M. Wilkens, Acta. Metal.
Vol. 32, No. 3. pp. 299–311.
1984. With permission]

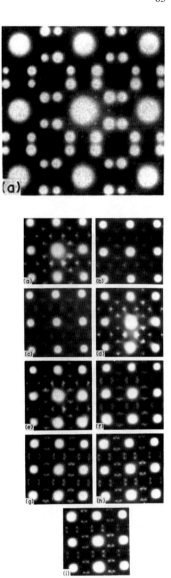

Fig. 5.38 Change of order
from SRO to LRO in Ni$_4$Mo
[after S. Banerjee, K. Urban
and M. Wilkens, Acta. Metal.
Vol. 32, No. 3. pp. 299–311.
1984. With permission]

diffraction pattern. There are more accurate models, one of which is due to Warren and Cowley [1967]. SRO leads to the appearance of diffuse intensity maxima in the reciprocal space in between the fundamental spots in the diffraction pattern as shown in Fig. 5.38a.

As the order parameter attains a value of 1, the diffuse intensity maxima sharpen to give rise to distinct superlattice reflections at those positions. See Fig. 5.37 for the locations of Superlattice reflections (as in Fig. 5.38i).

5.8 When the Material is Having Amorphous Structure

Amorphous materials completely lack any periodic arrangement of atoms. Since diffraction requires periodic set of scatterers so that constructive interference of the waves scattered by them takes place, we do not expect any systematic array of spots or rings from amorphous materials. The specific diffraction patterns from different amorphous materials show some differences, though there are many common features and regularity. The minimum distance between any pair of atoms and the maximum separation between them in solid state brings a correlation between them. Such correlation results in a ring pattern with the first ring being intense and diffuse, reflecting the correlation of first neighbour distances, followed by a weak diffuse ring as shown in Fig. 5.39. The number of observable rings in the pattern depends on the microscope conditions as well as the type of material, i.e. an amorphous alloy or amorphous carbon, etc.. The diffraction patterns shown in Fig. 5.39 illustrate this difference. The amorphous material in this case is a composite coating of Fe-based amorphous powders and nylon 11 developed by thermal spraying (high velocity oxy-fuel (HVOF) thermal spraying technique). The Nylon 11 powders turn into amorphous carbon particles in-situ within the Fe-based amorphous matrix. Hence distinct diffraction patterns could be seen in the two amorphous materials. When we observe an amorphous carbon film under a FEG illumination we see many diffuse rings in comparison to observation under a thermionic emission gun, due to the coherency of the beam in the former case. Similarly the diffuse rings from amorphous carbon, elemental amorphous thin films like Si and Ge, metallic glasses, oxide glasses, amorphous polymers and bulk metallic glasses show differences in the fine structure of the pattern besides a difference in the diameter of the first ring. Patterns from two such materials are illustrated in Fig. 5.40. The pattern in (a) is from an amorphous material obtained by mechanical milling of the constituent oxides of a borosilicate glass (Sastry et al. 2005). Note that the first and second rings (demarcated by blue rings) are both very diffuse when compared to those exhibited by well-known amorphous materials such as carbon (Fig. 5.39) and even more when compared to a bulk metallic glass, as given in (b). The pattern from a Bulk Metallic Glass (BMG) has

Fig. 5.39 DP of amorphous phase: [after Wei Wang, Cheng Zhang, Zhi-Wei Zhang, Yi-Cheng Li, Muhammad Yasir, Hai-Tao Wang and Lin Liu, Nature Scientific Reports, (2017)7: 4084. With permission]

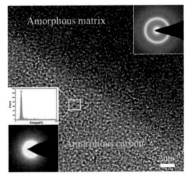

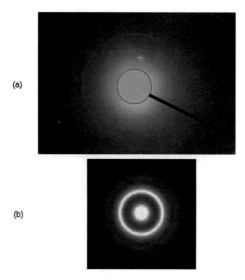

Fig. 5.40 DPs of different amorphous phases: **a** Mechanically milled mixture of oxides corresponding to the typical composition of a borosilicate glass. Note the very diffuse rings outside the blue rings. [after G. V. S. Sastry, S. Sudhir, Ankur Gupta, and B.S.Murty. "Milling of oxide blends–a possible clue to the mechanism of mechanical alloying", in Proc. "Seminar on Advanced X-ray Techniques in Research and Industry (XTRI-2003)", Ed. A. K. Singh, Capital Pub.Comp., 2005. With permission] **b** From a Bulk Metallic Glass

its own distinct features. The first ring is relatively sharp with a noticeable internal structure and is surrounded by a diffuse ring. Discussion on the existence of fine structure in the patterns from BMGs is beyond the scope here. Interested readers may consult relevant literature.

The diameter of the first ring reflects the average nearest neighbour separation distance in that glass, as mentioned earlier. From these patterns the nearest neighbour ordering correlation can be computed by plotting the radial distribution function $\rho(r)$. Inelastic scattering events also contribute to the diffracted intensity. The central point to be understood by us is that all amorphous materials are not alike and the diffraction patterns from them are similar on a gross scale but differ in their internal detail.

5.9 Convergent Beam Electron Diffraction

Convergent Beam Electron Diffraction (CBED) is a very powerful technique that can give a versatile crystallographic information about the material being studied. It derives the name from the convergent angle at which the electron beam is incident on the surface of a specimen. In the back focal plane of the objective such convergent incidence leads to diffracted beams that are divergent cones, instead of the parallel beams of conventional diffraction pattern. Though known since 1937 from the pioneering work of Möllenstedt and Kossel, the technique is not universally available on TEMs of all configurations and makes. By that we mean, the patterns cannot be obtained by a simple switching over from image mode to diffraction mode as we do in the case of a parallel beam diffraction. For that reason it is imperative to know the required configuration before we proceed to obtain CBED patterns.

5.9.1 The Instrument

There are two essential requirements: One, an extremely fine spot size at the specimen surface and second, the beam forming such a fine spot should be a cone with widest semivertex angle (note: we can form a fine spot even with a parallel beam, in which case it is called nano diffraction mode). Usually there are two-condenser lenses in any TEM, in some recent makes even three. The first condenser C1 forms a demagnified image of the gun crossover and thus determines the spot size. C1, therefore, should be strongly energised to form the smallest possible spot, closest to itself on the optic axis of the microscope. When the crossover is picked up by the second condenser, C2, it forms either a focussed spot or spreads the illumination evenly on the specimen surface. In the later latter case, the beam will be a parallel one. We require a focussed spot on the specimen surface with maximum convergence angle permitted by the C2 aperture. The ray diagram is shown in Fig. 5.41, by dotted lines.

The achieved convergence in this configuration is still inadequate. To improve upon this most microscopes are currently provided with an objective lens having split pole pieces. In a configuration of this type, the C2 lens is switched off while taking the CBED pattern (or is automatically off in this mode) and a highly convergent beam is focussed to a fine spot on the specimen surface by C3 (the objective pre-field); the angle of convergence being limited only by the C2 aperture (refer to the solid lines of the ray diagram, Fig. 5.41). Besides the above-mentioned requirement of condenser lens system, a clean microscope column with high vacuum, clean circular apertures and a plasma-cleaned flat specimen or a nearly-flat portion of a wedge shaped specimen are also essential for obtaining well-defined CBED patterns. Since we focus an intense beam on a small area, rapid build-up of carbon contamination cones takes place on either side of the specimen leading to diminished visibility of the pattern on the screen. Cooling of the specimen to liquid nitrogen temperatures is a remedial measure besides maintaining an oil-vapour-free column.

5.9.2 The Procedure

Align perfectly the C1 and C2 lenses with the help of gun crossover.

(i) Firstly, a zone axis pattern preferably with low index planes should be obtained by appropriately orienting the specimen. It is more convenient for further study,

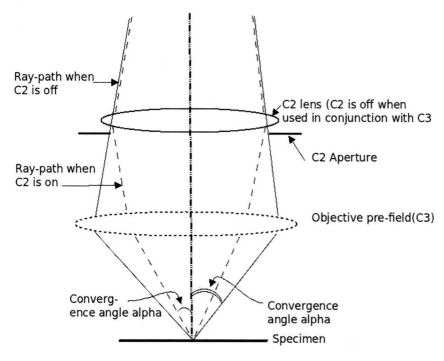

Fig. 5.41 Condenser configuration for CBED

    if this pattern could be located in any area of the specimen at 0,0 tilt position of the specimen holder.

(ii) Then we switch to image mode and centre the selected area under the beam. The area should be a reasonably flat portion of the specimen (recall the discussion on bend contours and thickness fringes to qualitatively assess the flatness of the selected region.).

(iii) We select the smallest of the available spot sizes on C1 (the numbers would be in reverse order of the spot sizes).

(iv) We then refocus the image using objective lens control (Caution: The image will be very dim due to the reduced spot size).

(v) The C2 needs to be focused at this stage (in recent models, it may get switched off automatically).

(vi) Switch to diffraction mode and remove the diffraction aperture to observe the full pattern on the screen. What we see is a convergent beam electron diffraction pattern. Figure 5.42 shows two such patterns, one recorded at a low camera length (a) and the other at a long camera length (b).

    The pattern shown in (a) consists of a central disc surrounded by many diffraction discs that are arranged so as to display the symmetry of the zone axis. The central

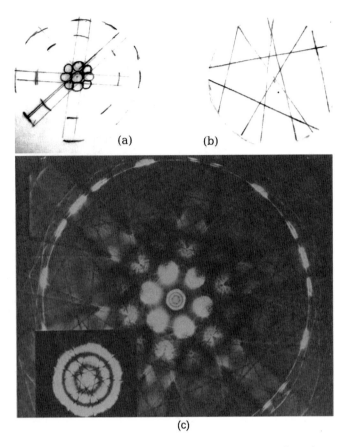

(a)   (b)

(c)

Fig. 5.42 Schematic CBED patterns at long **a** and short **b** camera lengths and **c** actual patterns (Short camera length pattern is shown as inset) [after Spence and Carpenter, Chap. 9 in J. J. Hren, J. I. Goldstein, D. C. Joy (eds.), Introduction to Analytical Electron Microscopy, Plenum Press, New York, 1979. With permission]

disc in CBED pattern is not referred to as transmitted disc but is either called 000 disc or direct beam for the reason that some diffracted beams (diffraction lines) also reach to locations inside the 000 disc (as in (b)). The reasons for their occurrence will be explained in a later section. By changing the diameter of the condenser aperture, we can either increase or decrease the diameter of the discs; and if angle of convergence β is greater than the Bragg angle the discs tend to overlap. Shifting of the condenser aperture position or beam tilt angle by manipulating the tilt controls that are provided in darkfield mode, has different effects on the pattern. Movement of C2 aperture shifts the location of the diffraction discs keeping the intensity profiles within them as they were, whereas beam tilt results in movement of diffraction discs along with intensity variations within them. The 000 disc remains unaltered.

The pattern is equivalent to that generated by the superposition of several patterns corresponding to the intersection of reciprocal lattice layers by the Ewald sphere orientations related to each of the multitudes of beams constituting the convergent cone of incident beam.

5.9.3 Measurements

Beam geometry of the CBED pattern and the corresponding ray diagrams need to be considered for deriving quantitative information from the pattern. The Fig. 5.43 shows a typical ray diagram for a systematics row.

The C2 aperture is completely filled with incoherent electrons and the conical beam subtends a semivertex angle of β on the specimen surface. Let point P be a location inside the cone from where a plane wave reaches the specimen surface. The un-deviated part of it reaches point 'T' in the 000 disc while the diffracted part reaches point 'S' in the corresponding diffraction disc **g** making an angle of $2\theta_{hkl}$ with the axis of the microscope, where θ_{hkl} is the Bragg angle for the **g** vector. Hence the measured distance 'r' between these two points on the screen (or on the recorded image taking into account appropriate magnification factor, if any)

$$r = L|\mathbf{g}|\lambda \tag{5.27}$$

Likewise, the diffraction condition is satisfied for every wave within the cone of incident beam. The Ewald spheres corresponding to each T can also be constructed.

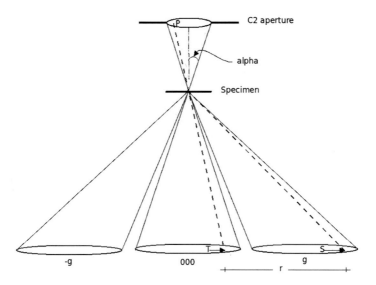

Fig. 5.43 Ray diagram for obtaining a systematics row

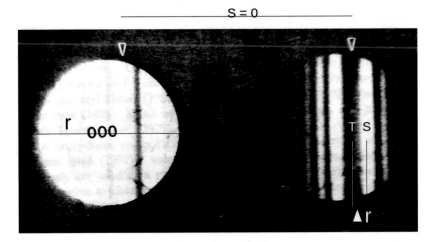

Fig. 5.44 Systematics row pattern [after J. C. H. Spence and J. M. Zuo, 'Electron Microdiffraction' Plenum Press, New York, 1992. With permission]

the optimum separation of the diffraction discs, i.e. non-overlap condition can be achieved when $\theta = \theta_{hkl}$. Neither the 000 disc nor any of the diffracted discs have uniform intensity. For example, corresponding to point T, there will be a minimum in the 000 disc and a maximum in the **g** disc at point S in the intensities. Since the discs are 2-D, these will appear as dark line at T in 000 disc and a bright band at S in the **g** disc in an actual pattern. In fact, this much of energy is transferred from direct beam to the diffracted beam. Thus, the diffracted discs in a systematics row, a row of discs **g**, 2**g**, 3**g**, etc., show bands of intensity maxima with a few corresponding dark bands in the 000disk. These bands are 2-D rocking curves that can be computed as per dynamical theory of diffraction (Spence and Zuo 1992; de Graef 2003; Reimer and Kohl 2008). Valuable information regarding extinction distance ξ_g, deviation parameter s_g and local thickness of the specimen can be obtained from a systematics row of an actual pattern (Fig. 5.44). We know that the length measured from edge of the direct disc to the opposite edge of adjoining diffraction disc is r and f om any point T to the corresponding point s; therefore, Δr is related to the deviation from Bragg condition s_g as

$$s_g \approx \left(\frac{\Delta r}{r}\right)g^2\lambda \tag{5.28}$$

Further, the deviation parameter s_i corresponding to an n_i^{th} minimum in intensity of the diffracted beam I_g, the extinction distance ξ_g and thickness are related according to dynamical theory (Spence and Zuo 1992) as

$$\left(\frac{s_i}{n_i}\right)^2 + \left(\frac{1}{n_i}\right)^2\left(\frac{1}{\xi_g}\right)^2 = \left(\frac{1}{t}\right)^2 \tag{5.29}$$

Fig. 5.45 A representative plot of s_i versus n_i for estimating $\left(\dfrac{1}{t}\right)^2$ and $\left(\dfrac{1}{\xi_g}\right)^2$ as slope

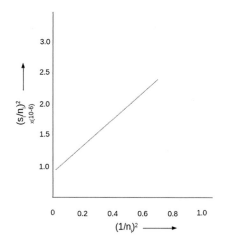

Therefore, by obtaining a systematics row and identifying various minima, one can plot $\left(\dfrac{S_i}{n_i}\right)^2$ against $\left(\dfrac{1}{n_i}\right)^2$ to get $\left(\dfrac{1}{t}\right)^2$ as intercept and $\left(\dfrac{1}{\xi_g}\right)^2$ as slope of the straight line. However, two points to be noted. One is related to the first minimum in the pattern (corresponding to S_i). It may be either inside or outside the Bragg condition. The other is related to the assignment of a value to n for $i = 1$. The convention is, if $t < \xi_g$, n is taken as 1 and when $\xi_g < t < 2\xi_g$, n=2 and so on. The method requires a strict two-beam condition which may not always be the case; and hence certain errors in the estimates. A least squares method of fitting the data is one solution to the problem (Fig. 5.45).

5.9.4 Zone Axis Pattern and Zero Disc Patterns

Symmetric zone axis patterns with multiple diffraction discs or only 000 disc containing several Higher Order Laue Zone (HOLZ) lines are very useful in extracting precise crystallographic information about the material being studied.

Crystal point groups: The symmetry elements that we observe in any zone axis pattern such as those shown in Fig. 5.46a or b only reflect 2-D point groups that should be related to the projection diffraction groups. They in turn need to be related to 31 Diffraction groups by observing whole pattern symmetry as well as that present in the Zero Order Laue Zone (ZOLZ) (Goodman 1975; Loretto 1984).

The identified Diffraction group from a set of zone axis patterns revealing HOLZ details is then related to the appropriate point group out of the 32 crystal point groups using the Table due to Buxton et al. (1976). From the point group thus determined, we can identify the space group of the crystal out of the possible 240 Space Groups of 3-D crystals by locating the presence of any forbidden reflections and observing the presence of dark bars – or + signs (called Gjønnes-Moodie or G-M lines) within them.

Lattice strain, change in lattice parameter etc.: The study of HOLZ lines in the 000 disc also offers valuable quantitative information due to their precise definition of location and width. These are dark lines in a bright background of the 000 disc and are also present in other diffraction discs. Their positions are sensitive to changes in accelerating voltage of the microscope, strain in the lattice caused by local changes in composition as per Vegard's law etc. These parameters are related to each other through the following equation (Spence and Zuo 1992):

$$-\frac{\Delta a}{a} = \frac{\Delta \theta}{\theta} = \frac{\Delta E_o}{2E_o} = \frac{\Delta g}{g} \approx \frac{2k_o\gamma}{g^2} \tag{5.30}$$

It arises from dynamical equation for intensity of a HOLZ line. Thus, the effect of strain on line position is strongly influenced by the variation in accelerating voltage and Bragg angle θ. HOLZ lines of highest order (with large h, k, l values) need to be selected to measure the shifts in line positions as a function of accelerating voltage. The simulated pattern in Fig. 5.47 shows line positions as a function of voltage when the accelerating voltage was changed from 100 kV (full lines) to 90 kV (dotted lines). The line positions themselves need to be calibrated and measured, instead the line intersections can be monitored. This method can be directly implemented on microscopes that provide continuously variable accelerating voltage by noting the ΔE_o required to bring back a chosen HOLZ line to its original position. The

Fig. 5.46 [100] Whole pattern symmetry and Bragg orientations along [200] and [220] showing symmetries of Cu-15at%Al **a** and [111] Whole pattern symmetry and central disc symmetries **b** of the same alloy. [after Marc de Graef, 'Introduction to Conventional Transmission Electron Microscopy', (Cambridge University Press), 2003. With permission]

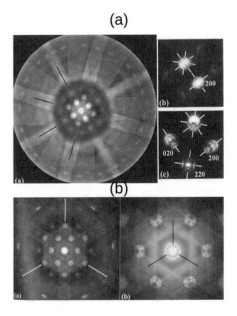

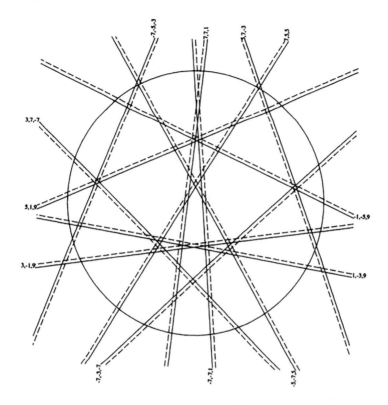

Fig. 5.47 Zero disc of CBED pattern showing HOLZ lines (a simulated pattern) [after J. C. H. Spence and J. M. Zuo, 'Electron Microdiffraction' Plenum Press, New York, 1992. With permission]

inaccuracies though are high, even when compared to X-ray method, the technique offers information local to a few tens of nanometers, say near a grain boundary. We also need to supplement the experimental data with computer simulations taking into account dynamical interaction. Besides that, the specimen needs to be kept at liquid nitrogen temperature as contrast of HOLZ lines is very sensitive to thermal vibrations. No dislocations should be present in the path of the beam. CBED patterns recorded with larger spot sizes reflect the whole crystal symmetry, i.e. its bounding faces, etc. Therefore, very small probe sizes that illuminate small flat regions of the specimen are required for better accuracy.

5.9.5 Defect Analysis

Burgers vectors of dislocations are determined typically by choosing three distinct **g** vectors for which a dislocation shows no contrast in the bright field image as discussed in Chap. 7. Three simultaneous equations are set up in $\mathbf{g \cdot b} = 0$ to solve for the $\mathbf{b}$ vector. Nevertheless, the tilting experiments to get to different $\mathbf{g}$ vectors satisfying $\mathbf{g \cdot b} = 0$ and $\mathbf{g \cdot b x u} = 0$ often are tedious and many a time residual contrast may persist in the image due to the anisotropy of the particular lattice. The Large Angle Convergent Beam Electron Diffraction (LACBED) proves to be a powerful tool by which the Burgers vector can be determined often from one single pattern.

LACBED technique involves raising the specimen from eucentric position towards C2 aperture so as to intersect the incident cone of electrons at a height δh away from the focussed spot. The ray diagram is depicted in Fig. 5.48. Under this condition a spot pattern is formed at the former specimen plane within the objective lens instead of at its back focal plane. When the diffraction lens is energised and the selected area aperture is focussed on to this plane, the resulting image on the screen will then be a superposition of bright field image and the CBED pattern. This is called LACBED pattern or Tanaka pattern (Cherns and Preston 1986; Tanaka and Kaneyama 1986; Tanaka et al.1988). By adjusting the strength of diffraction lens, we can either see only the 000 disc (Fig. 5.49) or the whole pattern. The area covered by the convergent beam on the specimen is much larger when compared to CBED. The diffraction discs under this condition display HOLZ lines together with image detail which is similar to bend contours and bend centres, if a dislocation line with line vector $\mathbf{u}$ is intersecting these lines, each HOLZ line gets split into subsidiary maxima of intensity where the HOLZ line exactly crosses the dislocation line. In such condition $\mathbf{g \cdot b}$ = n+1 for the dislocation under the HOLZ line, where $\mathbf{g}$ is the hkl corresponding to the particular HOLZ line, $\mathbf{b}$ the Burgers vector and n is the number of subsidiary maxima. Since we have a superposed image, a weak image of the underlying dislocation can also be seen in the 000 and diffraction discs. The sign of (n+1) needs to be decided according to the guidelines set up by Cherns and Preston (1986). As

Fig. 5.48 Ray diagram LACBED [Based on the diagram from Spence and Zuo 1992]

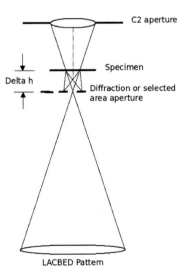

Fig. 5.49 LACBED pattern:
Bright field disc of
convergent beam pattern
recorded at 100 kV from an
Fe-30Ni-19Cr alloy. The
incident beam was along a
(114) axis and the beam
diameter was 200Å focussed
on a region of perfect crystal.
The arrows indicate the trace
of 110 mirror. [after
D.Cherns and A. R. Preston,
J. Elect. Micro. Tec. 13
(1989), 111–122. With
permission]

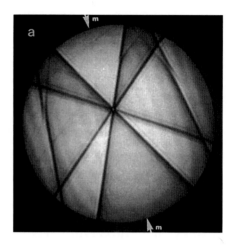

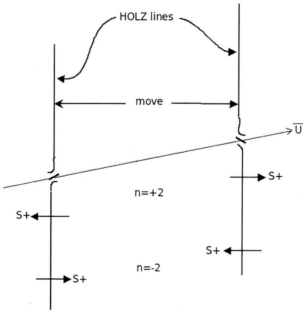

Fig. 5.50 HOLZ line splits; sign convention [Based on the diagram due to Cherns and Morniroli 1994]

shown in Fig. 5.50 the sign is +ve or −ve depending on the direction in which the HOLZ line splits bend vis-a-vis the positive direction of the deviation parameter s_g. An example taken from Tanaka et al. (1988), illustrates the method in the case of a dislocation in Si. The method can be applied to any type of dislocation, integral or partial and also can be extended to other defects such as stacking faults or interfaces. Considerable advancement in this direction has been made following the initial work by the Bristol group and Tanaka et al.

The technique also puts some restrictions upon its applicability, such as that the dislocation should not be close to either of the surfaces of the specimen and the diffracting conditions are away from the dynamical situation. The dislocation should be long enough to be crossed by at least three independent HOLZ lines.

5.10 Other Diffraction Techniques

There are a few other diffraction methods which offer precise crystallographic data that is often sought. Precession Electron Diffraction (PED), Rotation Electron Diffraction (RED), Reflection High Energy Electron Diffraction (RHEED), Electron Back Scattered Diffraction (EBSD) and so on. Of these, PED technique will be discussed here briefly and the discussion on EBSD will be taken up in the Chapter on Scanning Electron Microscope (SEM). EBSD is generally configured for an SEM though it is possible to obtain EBSD patterns in TEM as well.

5.10.1 Precession Electron Diffraction

The Bristol group invented the PED technique that offers precise crystallographic data similar to CBED (Vincent and Migley 1994).

The technique involves precessing the incident electron beam about the optic axis of the microscope at an angle of 1° to 3°. The incident beam can either be a parallel or convergent one. Both scan coils and descan coils are provided that control the precession and bring back the diffracted to the original position (as per Bragg angle) on the optic axis as depicted in Fig. 5.51 (Zou et al. 2011). The diffracted beam in a particular position of the precession is imaged as a fine spot. At this position the Ewald sphere samples only a few reciprocal lattice points and thus the diffraction condition remains kinematical only.

Certain hardware and software may be necessary to perform this experiment on any microscope. The digitally acquired patterns at each angle ω are put together to give rise to a complete pattern. The composed pattern hence has all the ZOLZ information together with HOLZ reflections or rings of reflections. The spots are very sharply defined in PED unlike those broad discs of CBED. Thus crystallographic symmetries can easily be deduced as shown in the example of 3-f, 6-f distinction in the case of a cubic mineral (refer to Fig. 5.23 of Mornirolli et al. 2010).

Fig. 5.51 Ray diagram for Precession Electron Diffraction [after Zhang, D. L., 2010, "3D Electron Crystallography-Real space reconstruction and reciprocal space tomography" Doctoral Thesis, Stockholm University, ISBN: 978-91-7447-044-4. With permission]

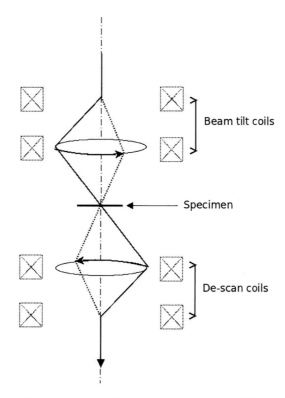

The resolution of PED is much higher compared to the normal selected area diffraction. In a 3° precession the outer most detectable spots of the ZOLZ corresponds to 0.25Å (Zhang et al. 2010). The resolution is better, the larger the precession angle ω, although overlap of Laue zones may occur at the same time. The patterns can be indexed in the same way as we did in the case of parallel beam electron diffraction patterns.

Exercises

Q1. The kinematical theory of diffraction contrast arrives at an expression for diffracted beam as $I_s = \frac{C^2(Sin^2\pi t s_z)}{(\pi s_z)^2}$. Discuss the conditions under which this expression is valid, particularly when the planes are diffracting at Bragg condition, by considering the various parameters of the expression.

Q2. When an electron beam passes along a zone axis a set of crystallographic planes, which are parallel to that direction, diffract the incident beam. While those planes are innumerable, we observe only a few reflections in the diffraction pattern on the screen. Explain the reasons for it.

Q3. Define and explain the following terms: (a) Extinction distance. (b) Deviation from exact Bragg Condition as against Kikuchi lines. (c) Multiple diffraction.

Q4. Compare the advantages of a convergent beam electron diffraction pattern with those of parallel beam electron diffraction pattern. Is the symmetric zone axis pattern in CBED in exact Bragg Condition? If yes, how do you explain the occurrence of dark and bright bands in the discs.

Q5. The two diffraction patterns (a) and (b) given below are recorded from the same single phase region of a specimen under two tilt conditions. It so happens that while (i) was recorded at 60 kV while (ii) was recorded after changing the accelerating voltage to 100 kV for obtaining better images. Index the diffraction patterns knowing that they belong to an f.c.c. crystal.

Chapter 6
Optical Microscopy

Optical microscopy forms core of the microstructural studies even today when several other techniques with higher resolution are available. That is because, a preliminary examination of the specimen in an optical microscope offers an overall phase distribution in the specimen, growth characteristics or thermal history of the specimen, and quantitative information such as grain size, volume fraction of particles in a two-phase alloy, particle size distribution and nanoscale surface roughness by special techniques. Concepts regarding resolving power of a microscope, super-resolution, ability of human eye to resolve the image and the theory of image formation due to Abbe and Rayleigh's criterion will be developed first.

Depth of field, depth of focus and minimum magnification achievable of an objective lens are other important characteristics to be known. The spherical lenses suffer from various aberrations which severely affect image quality and need to be corrected. Features of a typical metallurgical microscope are described in Sect. 6.4 with the help of a detailed as well as simple ray diagrams. Types of objectives that are either partially or fully corrected for aberrations are described along with examples. Oculars are essentially of two types, Ramsden and Huygenian; their features and specific use are also presented. Section 6.5 gives a description of polarised light microscopy and its applications along with a detailed discussion on birefringence. Section 6.6 deals with two types of interference microscopes, viz., interference-fringe microscope and interference contrast microscope or differential interference contrast microscope (DIC microscope) along with their respective applications.

6.1 Limit of Resolution and Resolving Power

The limit of resolution of a microscope is the smallest spacing between two object points that it is able to distinctly image and resolving power is the reciprocal of the same. It is attributed to the properties of an objective lens of the microscope since light reflected or transmitted by a specimen is first picked up by the objective lens for image formation. This parameter is proportional to the wavelength of the probing radiation used and reciprocal of the numerical aperture N.A. or the light gathering ability of the lens, expressed in other words, as shown in Fig. 6.1.

$$\Delta r_{min} \propto \frac{\lambda}{N.A.}$$

$$\Delta r_{min} = \frac{k\lambda}{\mu \sin \alpha} \tag{6.1}$$

where k is proportionality constant, λ the wavelength, μ the refractive index of the medium between the lens and object and α the semi-aperture angle subtended by the objective aperture at the specimen surface. Together with the refractive index, the Sinα term defines the light gathering ability of the objective lens and thus is termed as numerical aperture N.A. of the objective. k is a constant that can take values from 0.5 to 1 depending on the illumination, oblique, straight, etc. and generally has a value of 0.61. If the values of k and the denominator are so adjusted as to give a value of 1, then we notice that the minimum resolvable spacing would be of the order of wavelength of the probing radiation used.

Therefore, within the visible part of the electromagnetic spectrum, we have a limited choice from 700 to 400 nm. Thus, we are left with a handle on the denominator for improving the resolution. The medium between the specimen and objective can be changed from air, which has a refractive index of 1, to any other medium such as cedar oil which has a higher refractive index. This increase in μ leads to the bending of those light rays into the objective aperture which otherwise would have gone out of the objective aperture. Similarly, the Sine term can be maximised to 1, hypothetically, which means we have an objective lens of infinite diameter—an unrealistic proposition. However, there is a way out found for this method in what is

Fig. 6.1 Limit of resolution defined

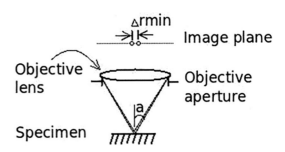

known as "Scanning Near-field Optical Microscope (SNOM)" which is described in Sect. 6.1.3. Thus, a routine optical microscope offers a limit of resolution of ≈ 200 nm. In Sect. 6.1.3, we will discuss some more novel concepts based on which this limit could be taken to much smaller values.

6.1.1 Abbe's Theory of Image Formation

The above-mentioned expression for limit of resolution (Eq. 6.1) is not based on simple geometrical optics, but has been arrived at by Earnst Abbe, way back in 1873, by considering diffraction effects. The object is considered as a parallel grating which diffracts the light incident on it. The objective lens of the microscope collects these rays and forms a diffraction pattern in its back focal plane. The diffraction spots further give rise to secondary wavelets that interfere to give rise to an image in the Gaussian image plane. Under direct axial illumination conditions of a coherent beam of light, Abbe proposed that the direct beam (un-deviated or transmitted beam) and at least one first-order diffracted beam are essential to form the image. The limit of resolution achieved under this condition is given by

$$\Delta r = \frac{\lambda}{N.A.}$$

He also realised that this limit can be improved further by considering oblique illumination (incoherent illumination) and Δ r can take the form

$$\Delta r = \frac{\lambda}{2N.A.}$$

Hence, larger the number of diffraction orders that are included in the objective aperture, smaller would be the resolved spacings: ideal image being the one where all diffraction orders are included.

6.1.2 Rayleigh Criterion

Rayleigh proposed a criterion, independent of Abbe, based on physiology of human vision. He proposed that two self-luminous objects (stars in his original proposal) i.e. point sources of incoherent light can be resolved by the human eye distinctly, provided that there is at least 80% drop in peak intensity on the line joining the centres of the two images. For this condition to be fulfilled, the peak in intensity of the second point source S2 should be coincident with the first minimum of the first source S1. This is depicted in Fig. 6.2.

Fig. 6.2 Rayleigh criterion
defined in one-dimensional
and two-dimensional Airy
discs

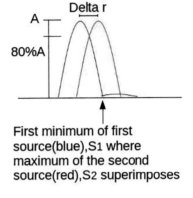

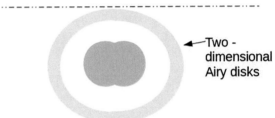

Intensity plots of both sources show subsidiary maxima of very weak intensity and in two dimensions they would be seen as weak diffuse rings of intensity surrounding the central disc of high intensity for each source, called Airy discs. Airy showed in 1835 that the image of a self-luminous object (star in his case) is not a point but a disc of intensity surrounded by several rings and is a diffraction pattern indeed. He arrived at the equation for the distance of the central maximum to the first minimum in the diffraction pattern as follows:

$$\Delta r_1 = \frac{0.61\lambda}{N.A.}$$

Thus, the Rayleigh criterion for resolution works out to be twice of this length. In terms of diffraction phenomenon, the minimum angle subtended by the two self-luminous objects which are seen just resolved, then works out to be

$$\theta_{(min)} = \frac{1.22\lambda}{d} \tag{6.2}$$

where d is the diameter of the aperture. It is for this reason that the resolution of an optical system is said to be 'diffraction limited'.

6.1.3 Possibilities of Extending Resolution Limits

It was realised in later years that Rayleigh criterion which was defined on the basis of human vision may not be absolute. It is argued that this minimum intensity or maximum contrast to perceive luminous objects may be tuned by some means. Several researchers defined the resolution limit in different ways that help in overcoming the reservation about Rayleigh criterion. Significant among them are (i) Scanning Near-Field Optical Microscopy (SNOM) and (ii) super-resolution microscopy of fluorescent molecules. We will briefly discuss the optical principles of these two techniques. Interested readers may refer to the comprehensive reviews on these developments Weisenburger and Sandoghdar (2015), Betzig et al. (1991) for further details.

SNOM: In the expression for limit of resolution (Eq. 6.1) due to Abbe, we have seen that one way of improving resolution is to collect diffracted rays of all orders for image formation (from the far field). For this to happen, we need to have either an infinitely larger aperture or go very nearer to the surface, i.e. near field. However, in the latter case, we cannot use the light of same wavelength as for the far-field microscope (normal optical microscope, where the distance between sample surface and the objective lens is 50 to 100 times more than the wavelength of light used) as realised by Synge in 1928.

He took advantage of non-propagating evanescent waves generated at the surface when light is shined on it. Basic design comprises a very fine light source shining out of a metal coated tip of a glass fibre positioned near a 20 nm diameter aperture drilled in a metal foil. The metal foil in turn is positioned very close to the surface of the specimen ($d \ll \lambda$). The evanescent fields containing high spatial frequency information of the surface die down within a short length of one wavelength of radiation used. His idea was later put to use by Ash and Nicholls (1972) where they used microwave radiation to obtain a resolution of $\lambda/6$. Dieter Poll of IBM, Switzerland and Aaron Lewis of Cornell University developed the modern version and used light from visible spectrum. Many designs are available for the light source of which the Aperture SNOM will be discussed here. It consists of a metallised glass fibre tip that is cladded in ceramic and metallised outside for robustness. This light source has a sub-wavelength aperture at the bottom and is rastered or vibrated close to the surface of the specimen at a distance $d \ll \lambda$. The evanescent fields so generated are collected either at the far field or by the same aperture (see Fig. 6.3). Resolutions up to 50 nm are achieved in this way using ordinary light. The technique

Fig. 6.3 A typical aperture
SNOM

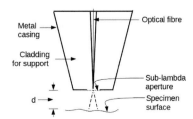

proved to be extremely useful in cell biology studies and also in semiconductor device characterisation. Nevertheless, it has certain experimental drawbacks such as difficulty of reaching down to specimen surface close to 1 nm (this is obviated with the development of Atom Force Microscopy, etc.), poor signal-to-noise ratio when evanescent fields are sensed by the same aperture.

Super-resolution: Many far-field optical microscopy techniques have been developed to take the resolution limits well beyond those achievable by near-field techniques such as SNOM, even to the extent of single molecule imaging (single molecule localisation techniques). These techniques were developed on the basis of fluorescent properties of some biological molecules.

(i) Simulation Emission Depletion (STED) technique: Steffan Hell conceived the idea of suppressing the fluorescent radiation at the outer regions of the beam that is emitted by a fluorescent molecule upon excitation by a laser. Another laser (of different wavelength) is used for the suppression, thereby improving the resolution. (ii) Single molecule detection: As early as 1970, the idea of being able to locate a single fluorescent molecule embedded in a non-fluorescent matrix though was known, it was only in 1993 that Betzig could demonstrate the feasibility down to about 50 nm.

Various other techniques such as Photoactivation Localisation Microscopy (PALM), Stochastic Image Reconstruction Microscopy (STORM) and the like, not only attempt to take the resolution limits of optical microscopes to a few nanometers but also, importantly, without taking help of any fluorescent behaviour of the constituent molecules. Considerable statistical analysis of the digital data of the pixels of the acquiring CCD camera is involved in these methods. Optical Diffraction Tomography (ODT) and Surface-Enhanced Raman Scattering (SERS) are steps towards taking the resolution to nanometer levels.

The importance of these super-resolution techniques can be gauged from the fact that 2014 Nobel Prize for Chemistry was shared by Steffan Hell, for developing STED, Eric Betzig and W. E. Moerner for independently developing single molecule microscopy.

6.2   Magnification, Depth of Focus and Depth of Field

Magnification can be variously defined with respect to an objective lens as shown in Fig. 6.4:

$$M = \frac{Image\ distance,\ v}{object\ distance,\ u}$$

or

$$M = \frac{size\ of\ the\ image,\ y}{size\ of\ the\ object,\ x}$$

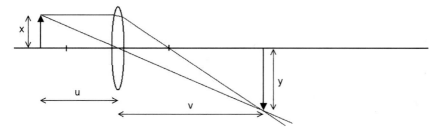

Fig. 6.4 Definition of magnification

In a microscope the image formed by the objective is further magnified by an ocular. Therefore, the total magnification achieved by the microscope would be a product of the two.

$$\text{i.e. } M_{micro} = M_{objective} \times M_{ocular}$$

There can be further modifications to this definition as we see in a later section on compound microscope.

Minimum total magnification, useful magnification and empty magnification: We have already seen that the minimum separation between two points on an object that can be distinctly imaged by an objective lens is its limit of resolution, Δr given by

$$\Delta r = \frac{0.61\lambda}{N.A.}$$

The two object points or sources of scattering of light S1 and S2 that are separated by a distance Δr on the specimen are imaged by the objective lens of the microscope as S1' and S2' separated by a spacing of $\Delta r'$. If M_{ob} is the magnifying power of the objective then $\Delta r'$ is given by

$$\Delta r' = \Delta r M_{ob} = \frac{0.61\lambda}{N.A.} M_{ob}$$

But, this image, though resolved, (see Fig. 6.5) cannot be seen by the eye. We know that human eye has a limit of resolution of 0.1 mm (Δr_e) at a distance of 250 mm or 1.5' of arc. Even if we choose middle of the visible spectrum of light that has a wavelength of 540 nm, by no enhancement of the magnification of the objective lens alone can we render the image visible to the human eye. Therefore, an ocular of suitable magnifying power is necessary for visualising a distinct image.

Fig. 6.5 Minimum magnification to be maintained by the microscope to visualise the resolved image

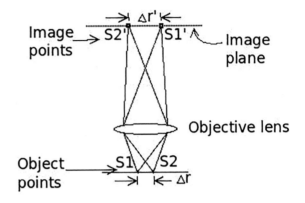

$$M_{oc}(min)\Delta r' = \Delta r_e$$

or

$$M_{oc}(min)M_{ob}\Delta r = \Delta r_e$$

If minimum magnification $= M_{(min)}$

Then $M_{(min)}\Delta r = \Delta r_e$

or

$$M_{(min)} = \frac{\Delta r_e}{\Delta r}$$

For the chosen λ and Δr_e of 0.15 mm this works out to be $M_{(min)} = 455$ N.A.

and

$M_{(max)} = 1000$ N.A.(corresponding to λ for deep UV)

Hence, we can define $M_{(useful)}$ as

$$455 \text{ N.A.} \leq M_{(useful)} \leq 1000 \text{ N.A.}$$

Any magnification achieved beyond this is termed as 'empty magnification' because the image cannot give any further details, i.e. the resolution cannot be enhanced further by a mere increase of magnification.

Depth of focus: It is the ability of a microscope to render a sharply defined image over a length Δv about the Gaussian image plane on the optic axis. It is an important parameter of the microscope that gives us leverage in placing a screen or a camera to capture a sharply defined image (see Fig. 6.6).

Depth of field: It refers to the object field and expresses the ability of a microscope to simultaneously bring into focus, object points at different levels of depth on the object, in the image plane. If Δu is the spread of depth levels around a mean object

Fig. 6.6 Definition of depth of focus

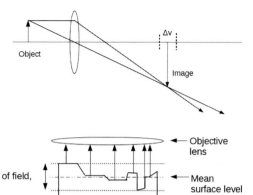

Fig. 6.7 Definition of depth of field

plane, then Δu can be expressed as follows:

$$\Delta u \approx \frac{\mu A}{(N.A.)^2}$$

where μ is the refractive index of the in-between medium, λ the wavelength of light used for imaging and N.A. is the numerical aperture of the objective (see Fig. 6.7).

Long focal length converging lenses have larger depth of field. One way of providing larger depth of field for optical microscopes is to build in stereo vision just like that of humans (animals in general). A stereo microscope is routinely used to examine bulk un-polished specimens whose surfaces are left pristine and non-levelled. Such surface characterisation is vital to materials study.

6.3 Lens Aberrations

Due to the geometry inherent to them, the lenses, both convex and concave, suffer from many aberrations. The principal two types that severely affect the image quality are spherical and chromatic aberrations.

6.3.1 Spherical Aberration

Due to spherical aberration of the lens, the rays which travel close to the optic axis of the lens get focussed farther on the optic axis when compared to the parallel rays that are travelling along the peripheral regions of the convex lens which get focussed closer to the lens. The same is true for a concave lens, where the marginal rays appear to have focussed closer to the lens while the axial rays appear to have focussed farther on the optic axis. This differential focussing is due to the spherical curvature of the

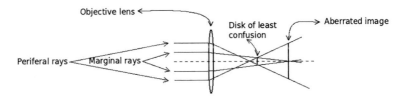

Fig. 6.8 Spherical aberration

lens and is to some extent independent of the lens material and occurs even in thin lenses.

Since curvature is constant from centre to the edge of the lens, spherical aberration persists even among two axial rays, however close they may be to the axis of the lens. Therefore, one ingenious way of overcoming this aberration is to *gradually change the curvature of the convex lens from positive at the centre to negative at the edge of the lens*. Such lenses are already put to use by some manufacturers in special equipment. Common practice is to place a concave lens close to or cement with a convex lens to minimise or eliminate spherical aberration, as they refract light rays in opposite directions. On the image side, spherical aberration leads to spreading of the image of a point object into a disc, the diameter of which is the least at the location shown in Fig. 6.8. It is called 'disc of least confusion'. This aberration is actually termed as 'longitudinal axis spherical aberration'. On the image plane it is called 'lateral spherical aberration'. Hence, spherical aberration severely effects resolving power of the microscope.

6.3.2 Chromatic Aberration

For similar reasons as mentioned above, the light rays of longer wavelength, i.e. closer to red colour, are focussed farther on the axis, while those of shorter wavelength are focussed nearer on the axis resulting in a colour separated image. This is a severe aberration while using white light in the microscope for imaging. Correction of chromatic aberration, once again, is possible by combining convex and concave lenses of low- and high-dispersion materials such as corning glass for convex and flint for concave lens.

6.3.3 Astigmatism and Coma

Astigmatism arises as a result of differential focussing action of the lens along **x**- and **y**-axes depicted in Fig. 6.9. This is on-axis astigmatism due to manufacturing defects in the lens. Astigmatism arises often in otherwise defect-free lenses due to the rays that emerge from object points very much away from the optic axis of the microscope. They get focussed at two different focal points, one for x-axis and the other for y-axis as shown in Fig. 6.9.

Once again, a disc of least confusion is formed. The net result is blurring of a disc image corresponding to an object point.

Coma is a severe distortion of the image formed by off-axis rays that reach different points in image plane as shown in Fig. 6.10.

This leads to the formation of a comet-like image with an intense dot corresponding to the point object followed by a tail of dots formed by other rays.

6.3.4 Curvature of Image Plane

It is also called field curvature. Curvature of the image plane takes place since the central rays from a square object, say, get focussed farther on the optic axis while the rays coming from the peripheral regions get focussed closer. As a result, the outer regions of the square object appear out-of-focus compared to the central regions (see Fig. 6.11).

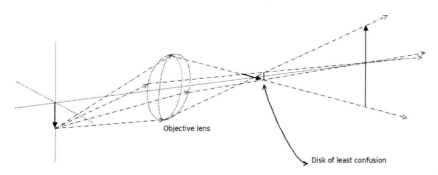

Fig. 6.9 Astigmatism

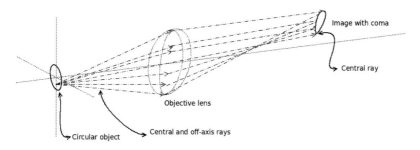

Fig. 6.10 Coma

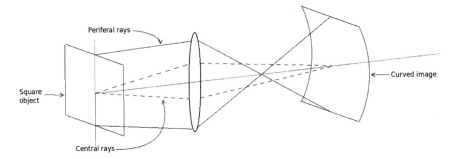

Fig. 6.11 Curvature

6.3.5   Distortions

Pin cushion and barrel distortions also occur due to differential focussing whereby a square object is imaged as a pin-cushion-shaped object or as a barrel-shaped with diagonal extensions as shown in Fig. 6.12.

6.4   A Compound Metallurgical Microscope

A compound microscope consists of at least one objective and one ocular, but may contain some tube lenses also depending on the sophistication of the microscope (see Fig. 6.13). A metallurgical version of it has a beam splitter to deflect light on to the non-transparent specimen and also transmit the light from the objective to the ocular. Most often, the currently available routine as well as research microscopes are equipped to view both transparent and non-transparent specimens. Light from a condenser or collimator is incident on a half-silvered mirror or a right-angled glass prism which reflects it through the objective lens on to the specimen. The illumination systems currently in use are described in Appendix D.

Fig. 6.12 Pin cushion or barrel distortions

The light reflected by the specimen detail is again collected by the objective and is brought to focus at the entry aperture of the ocular. The real image formed at this stage is magnified slightly and is inverted. This image acts as an object for the ocular and together with the lens of the eye forms an inverted real image on the retina, which is upright with respect to the object. The optic nerve behind the retina carries this image to the brain which re-interprets it as a virtual image formed at a distance of 250 mm from the eye. This virtual image is highly magnified, inverted with respect to the object and perceived to be formed behind the object so as to get an impression of viewing a magnified object. The ray tracing in dashed lines shows this image in Fig. 6.13. Notice that the objective and ocular are not single lenses but an assembly of lenses so designed as to minimise the aberrations discussed in the earlier section. The forward and back focal planes of the assembly can also be worked out. The equivalent simple schematic diagram is shown in Fig. 6.14.

In both cases, the object, i.e. the specimen is placed slightly outside the forward focal plane of the objective. The gap between coverslip of the objective lens and specimen is called working distance, which one desires to maximise to accommodate heating or cooling stages, etc. On the contrary, if the object is placed exactly at the forward focal plane of the objective then image is formed at infinity by the parallel rays sent out by the objective. Most of the modern microscopes adopt this design which offers a large space in the microscope tube from where the image can be

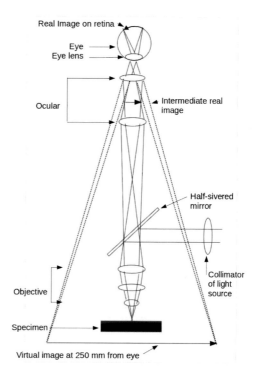

Fig. 6.13 Compound microscope

Fig. 6.14 A simple ray
diagram of the microscope

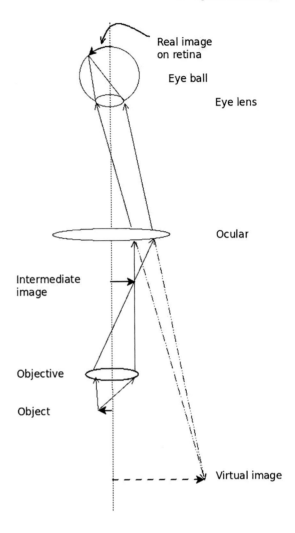

diverted to more than one path, or accommodate special devices such as polarisers,
DIC prisms, etc. This space is called infinity space and does not disturb the image
quality or focal points of the microscope provided the auxiliary components added
have no 'lensing' action. However, the first intermediate image is formed with the
help of a tube lens only as shown in Fig. 6.15.

The total achieved magnification would be

$M_{total} = M_{obj} x M_{oc} x M_t$ where M_t is the magnification of the tube lens system.

Another feature of a modern compound microscope is that it provides multiple
objective lenses having different magnifying powers on a single nose piece. Such
provision also requires the image to be in focus when one objective is replaced by
another by the rotation of the nose piece. For this purpose, the objective lenses have

Fig. 6.15 A simple ray
diagram of the microscope

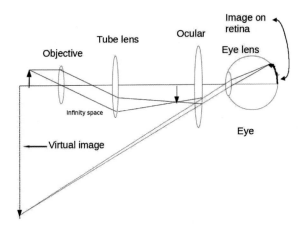

extendable bodies by turning one part of the objective metal casing on threads of the
other. This is called parfocalisation of the microscope.

6.4.1   Objectives and Oculars

Objectives of modern microscopes are never single lenses but often are a combination of convex, concave and plano-concave lenses. They are also coated with films to
either correct some of the aberrations or to improve the numerical aperture. From the
application point of view objectives can be categorised as follows: (i) Achromatic.
(ii) Plan achromatic. (iii) Apochromatic. (iv) Fluorite.

Achromatic lenses are corrected for chromatic aberration for the wavelengths corresponding to red and blue light. Spherical aberration is also corrected for these
objectives in the mid-wavelength of the visible spectrum. They perform well under
white light illumination and are cost-effective.

Fluorite or semiapochromatic are made of CaF_2 or Fluorspar mineral or out of
synthetic Lanthanum fluorite crystal. They are highly transparent, least colour dispersive and hence are suitable for fluorescent microscopy. They can also be used for
hot-stage microscopy.

Apochromatic objectives are highly corrected for chromatic aberration and thus are
very expensive. They are best suited for fluorescence microscopy besides differential
interference contrast and polarised light microscopy. The problem of field curvature
with these lenses is corrected in plan-apochromatic lenses. A commercial objective
lens has much of the data related to the lens inscribed on the outer casing, either in
codified form or in numbers. Figure 6.16 shows a typical lens.

Oculars are essentially of two types, viz., Ramsden and Huygenian types shown in
Figs. 6.17 and 6.18, respectively. Ramsden-type ocular consists of two plano-convex
achromatic lenses that are mounted with their curved surfaces facing each other. The

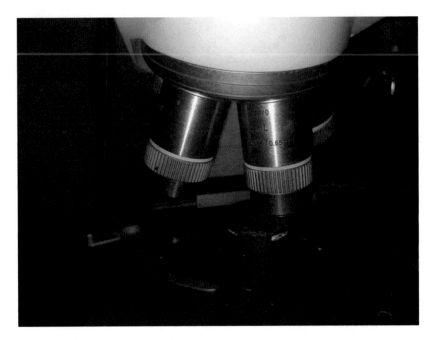

Fig. 6.16 A typical lens with data inscribed on it

field aperture in this design is placed outside the field lens towards the objective and the image forms just outside of the ocular. This design facilitates placing of any measuring graticule on the field aperture. For focussing the graticule on the image plane, the distance between eye lens and field lens is made adjustable in Ramsden ocular.

The Huygenian-type ocular has its field aperture placed between the eye lens and field lens. The convex surfaces of the two lenses face towards the objective and the optical distances are adjusted such that the image is formed within the barrel between the eye lens and the field lens.

Other designs: The eyeball is required to be placed close to the exit pupil of the ocular in the above designs, which is strenuous for longtime viewing, particularly for people with eyeglasses. Therefore, certain special designs provide the Ramsden disc (a disc of intensity seen floating around the exit pupil) to be formed well outside the exit pupil and make prolonged viewing comfortable.

Fig. 6.17 Ramsden-type
ocular

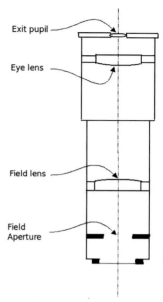

Fig. 6.18 Huygenian-type
ocular

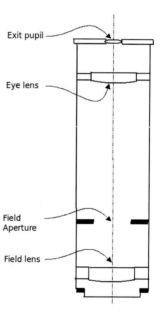

6.5 Polarised Light Microscopy

6.5.1 Birefringence

Natural white light is a mixture of many colours of different wavelengths and travels with its electrical vector vibrating randomly in all directions in a plane that is perpendicular to its direction of propagation in vacuum. When the light travels through isotropic media such as air, glass, cubic crystalline materials and amorphous polymers also, it does the same. Isotropic media are those materials that possess the same physical properties, e.g. refractive index in all internal directions. On the contrary, anisotropic materials offer different refractive indices for different propagation directions. Usually this is only one direction for many crystals while some exhibit anisotropy in properties in more than one crystallographic direction. Ordinary light can get polarised under certain reflection conditions also. When an unpolarised light beam is incident on the surface of glass or water (both of which are isotropic media) at a critical angle called Brewster angle a polarised light component gets reflected from the surface while the remaining component gets refracted into the medium. The reflected ray gets polarised along a direction that is parallel to the surface of the medium. Refer to Fig. 6.19.

One of the early realisations of polarisation effect in crystals is in rhombohedral calcite crystal. When light is passed along the optic axis of a freshly cleaved calcite crystal, it propagates in the usual way as in an isotropic medium. Instead, if the light enters the crystal in any other direction, it gets split into two rays that travel in different directions with different velocities. While doing so, the electrical vector of

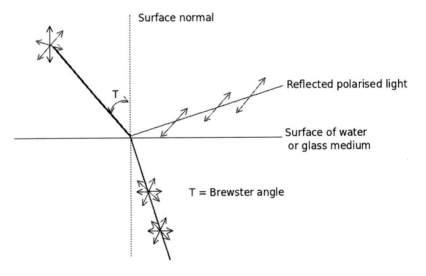

Fig. 6.19 Polarisation of reflected light

the light rays interacts with the electrical field of the lattice molecules and align in one direction. The electrical vectors of the two beams would be mutually perpendicular and their velocities would also be different. The slow ray is called ordinary ray and the fast one is called extraordinary ray. The term ordinary ray may be misleading at times, but should be remembered that it is also polarised. The retardation of the slow ray 'R' is related to the difference in refractive indices of the anisotropic medium in the two crystallographic directions as follows:

$R = tx(|n_e - n_o|)$ where 't' is the thickness of the crystal.

Note that the difference in refractive indices could be either positive or negative. This phenomenon is called birefringence. The relative velocities of the o-ray and e-ray can be understood by constructing the ellipsoid of refractive indices. From the ellipsoids, it can be seen that e-ray is faster for negative birefringence and is slower for positive birefringent crystals. Calcite crystal has a very high degree of birefringence. In practice, if a freshly cleaved crystal of calcite is kept on any printed paper, the letters on it would be seen as double images. Referring to Fig. 6.20, one can see that any printed letter on the paper, such as 'A', would appear in two images, one formed by o-ray and the other by e-ray (as in (b)), provided the incident light direction is neither $\perp$ nor $\parallel$ to the polarising direction 'P' of the crystal. If the crystal is rotated on the paper, one of the images stays stationary while the other precesses it. The stationary one is due to o-ray and the precessing one is due to e-ray. We can place a polariser film on the crystal and notice that when the polariser axis is perpendicular to the ordinary ray (Fig. 6.20c), the bottom 'A' disappears and when the polariser axis becomes parallel to that of the o-ray, the top image vanishes (Fig. 6.20a). The o- and e-rays emerge on the surface as two separate beams, as they travel in different directions in the crystal. However, when they happen to travel in the same direction their planes of vibration will still be mutually perpendicular. Such possibility exists when a polarised light beam is incident normally on a birefringent crystal surface, the o- and e-rays travel in the same direction but with a phase lag. They still cannot interfere with each other as their planes of vibration are mutually orthogonal. When the phase lag is either λ or $\frac{\lambda}{2}$ they give rise to a wavefront that is linearly polarised, while their superposition gives rise to a circularly polarised wavefront when the phase lag happens to be either $\frac{\lambda}{4}$ or $\frac{3\lambda}{4}$. If the phase lag happens to be any other fraction of

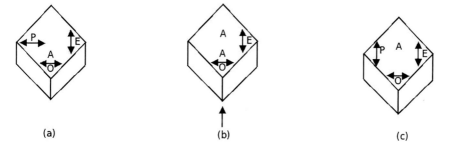

(a)   (b)   (c)

Fig. 6.20 Polarisation of light entering a calcite crystal in different directions (schematic)

λ we get an elliptically polarised light. The phase difference in turn is dependent on the difference of refractive indices offered by the medium for the two rays.

Classification: The above-discussed class of materials, both positive and negative birefringent, represent intrinsic birefringent materials. Birefringence can also occur in materials or structures due to physical conditions that cause anisotropy in properties. Natural materials such as muscle fibres, chromosomes and synthetic long-chain polymers, composites, etc. exhibit anisotropic refractive behaviour called *structural birefringence*. Physical stress on otherwise isotropic materials such as glass lenses, stretched films and fibres also leads to birefringence called *stress birefringence*. Strain gradients also may be present in the material which may be elastic, e.g. large casings or plastic also lead to *strain birefringence*.

6.5.2   The Microscope

Birefringence property of either naturally occurring crystals such as calcite and quartz or synthetic polymer films with known axis of polarisation is put to use to obtain contrast from anisotropic phases in isotropic matrix. The microscope is fitted two polarising crystals/thin films with mutually perpendicular polarising directions to begin with.

In reflection configuration, one of the crystals is kept above the objective in the incident light path which polarises the light received from the source, parallel to its polarising axis as shown in Fig. 6.21. This polarised light is shined through the objective on the specimen surface. Constituent phases of the specimen reflect light as per their birefringence properties; isotropic phases, e.g. ferrite or austenite phase, return the polarised light without any change in its polarisation direction, while any anisotropic phase (e.g. inclusions such as sulphides) present in it will return the light with a changed polarisation direction as per their birefringence properties. The objective lens picks up these beams and forms an image at the entrance pupil of the ocular after passing through analyser. The analyser in its initial condition stops all the beams as their planes of polarisation are not parallel to its own. For this reason, the analyser is mounted in a housing that can be rotated, by the operation of which the analyser's plane of polarisation can be made parallel to the plane of polarisation of the beams reflected by the anisotropic phase(s). The analyser in that orientation admits light to reach the ocular. In this condition, we will observe bright images of the birefringent phase(s) in a dark background corresponding to the matrix phase. The anisotropic phases referred to here are sulphide inclusions in steels which are detrimental to the mechanical strength of the steel. Polarised light microscopy is extensively used for reporting the inclusion count of a steel. We may also notice the possibility of keeping both the polarisers in parallel to begin with. In such a condition, the isotropic matrix phase would appear bright in the image as it does not change the polarisation direction of the incident polarised beam and the anisotropic inclusions would appear dark. We prefer the bright contrast of the anisotropic inclusions though.

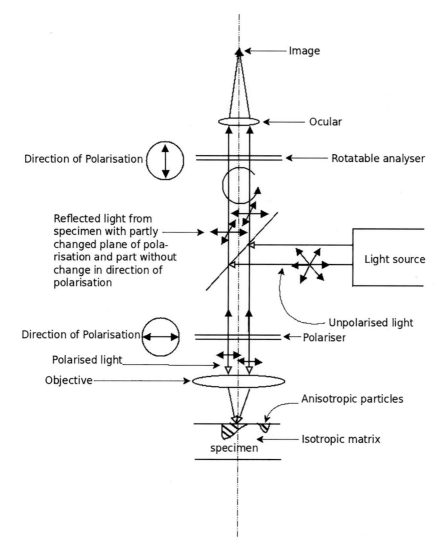

Fig. 6.21 Polarised light microscope

Applications: Besides the above-cited example, polarised light microscopy is useful in the study of wear tracks on a bearing surface made of an isotropic alloy. Even the glass lenses used in polarised light microscope need to be examined in polarised light as they may develop stress birefringence due to improper mounting. Though, glass by itself is isotropic, stress on it can lead to polarisation of the charges of its ions leading to changes in refractive index in some directions.

6.6   Interference Microscopy

Two coherent light rays reflected by the specimen and a reference plate would constructively interfere if their phase difference is $n2\pi$ where 'n' takes 0, 1, 2.... integer values. For all other values of phase difference, there would be attenuation of the intensity (square of the amplitude), maximum being at π or $\frac{\lambda}{2}$. Therefore, an interference pattern of dark and bright fringes is generated upon superposition. The topography of specimen surface, that is, height differences that might have occurred due to any reason, can be interpreted either directly from the (i) interference pattern or from (ii) an image constructed from the interference pattern. The first technique is called interference-fringe microscopy and the second one is phase contrast microscopy. A better alternative to the latter is Differential Interference Contrast (DIC) microscopy. The scale of contrast generated by DIC is much wider compared to the conventional phase contrast microscopy and hence will only be discussed here. Interested readers may refer to Pluta (1988) for details of phase contrast microscopy.

6.6.1   Interference-Fringe Microscopy

This microscope can be viewed as a combination of a Michelson interferometer and an optical microscope. It is schematically presented in Fig. 6.22. The microscope has a coherent source of monochromatic radiation that sends it to a beam-splitting prism. The beam splitter sends one beam straight to a fixed reference mirror through an objective and the other in a perpendicular direction towards a specimen, whose surface roughness needs to be characterised, once again through an objective that is matching with the reference one. A compensator is also placed in this path to make the distances travelled by the reference beam and specimen beam exactly identical.

The two matched objectives return two images that are superimposed to form an interference-fringe image at the ocular which magnifies it further. In the final magnified image (to be called diffraction pattern?) of the interference pattern, any vertical height differences on the specimen surface, such as growth steps on a single crystal or machining defects on a finished component, manifest themselves as shifts in the fringes at that location. The height 'h' is given by

$$h = \frac{d}{l} \times \frac{\lambda}{2}$$

where d is the shift in fringe spacing in units of fringe spacing, in other words a normalised value, and λ is the wavelength of the monochromatic light used. Obviously, the resolution of fringe spacing depends on the limit of resolution of the microscope and its magnifying power. The achievable resolution in this case of two-beam interference is about 25 nm. It has many applications in metallurgical and materials fields such as estimation of surface roughness of precision-machined components, directionally solidified alloys, semiconducting devices and thickness measurement

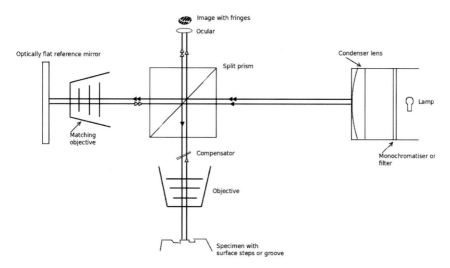

Fig. 6.22   Interference-fringe microscope

Fig. 6.23   Multiple-beam
interference-fringe
microscope

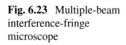

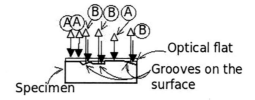

of multi-layer thin films. The resolution can be further enhanced by adopting a multi-beam interference. In this design, an optically flat transparent plate having almost the same reflectivity as that of the specimen is kept on top of the specimen surface as shown in Fig. 6.23.

This set-up is called multiple-beam interference-fringe microscope and is somewhat difficult to perform. The beam incident at 'B' passes through a wedge-shaped gap and undergoes interference multiple times and the fringes appear to be shifted with respect regions such as 'a'. The resolution in the multi-beam case can be as high as 1 nm.

6.6.2   *Interference Contrast Microscopy or DIC Microscopy*

Here, the interference pattern formed in the previous case is taken forward to form an image. A further innovation is to make the ordinary and extraordinary rays emerging from a birefringent crystal, which normally vibrate in mutually perpendicular directions to interfere with the help of an ingenious device called Wollaston prism.

Fig. 6.24 Ray diagram for a
DIC microscope

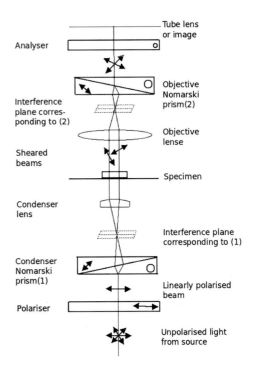

The design was conceived by Nomarski. The microscope converts the gradients in
optical path lengths, traversed by the polarised beams in the object space, to contrast in the image. In this way, this technique is similar to phase contrast microscopy
without any phase plate or the drawbacks of that technique. As the gradients and
not the actual path lengths are used for achieving contrast, the technique is called
Differential Interference Contrast (DIC) microscopy.

The principle: The optical path length is the product of refractive index and local
thickness of the transparent sample. Therefore, if a pair of o- and e-rays of polarised
light, which have an initial known phase difference, experience local changes in
refractive index or thickness or both while travelling through the transparent specimen
(or in reflection in case of opaque specimens) with a consequent difference in path
length and phase difference. Each pair of rays will have this difference and there
will be many such pairs delivered by the specimen. It is the recombination of these
pairs that generates contrast, which is brought about by another matching prism and
a suitably oriented analyser.

The microscope: Various components of the microscope are schematically shown in
Fig. 6.24. A parallel beam of white light or monochromatic light from a light source
is linearly polarised by a polariser whose polarising direction is fixed in left-right
or east-west direction. In this microscope, it is important to orient the polarising
directions or optic axes of all the conjugate components so as to maintain extinction
condition (zero intensity at ocular) as a standard state when no specimen is mounted.

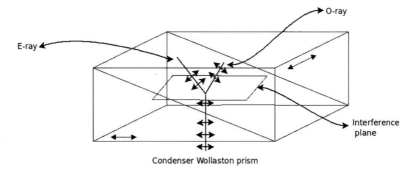

Fig. 6.25   Condenser Wollaston prism

The polarised light then meets first of the two Wollaston prisms which acts as a beam splitter. The two components of the prism are two cut wedge pieces of the same quartz crystal (or calcite) which are rejoined together so as to have the optic axis of the bottom wedge parallel to the optic axis of the polariser and that of the top wedge perpendicular to it. In this orientation, the linearly polarised light from the polariser gets split or sheared into o- and e-rays at the interference plane. The typical shear distance is almost 0.2–2 μm, which is an important parameter in deciding the contrast and resolution in the image, the closer they are the higher is the resolution but with some loss of contrast. Note that in recent designs the first prism, also termed condenser prism as shown in Fig. 6.25, is also a modified Wollaston prism. The emerging o- and e-rays are collected by the condenser lens that focusses the pair of rays as a parallel beam on the specimen. Thus, each object point in the specimen which is a phase object is illuminated by this pair. Optical path differences may be introduced by the specimen due to the reasons already mentioned. Beams of o- and e-ray pairs reach the objective lens, which is matched to the objective Wollaston prism and are brought to focus, i.e. made to interfere at the interference plane of the prism. By this we mean that the two rays of a pair are made to vibrate in the same plane so that they can interfere with each other as per their phase differences. For this it is more convenient to have the interference plane outside the Wollaston prism and located at the back focal plane of the objective lens. The older, unmodified Wollaston prism has this interference plane within the prism (as shown for the condenser Wollaston prism) and hence it is modified in the design of Nomarski prism. The optic axis of the bottom wedge of the split prism is oriented at 45° to the outer surface of the prism. Such a prism is called modified Wollaston prism or Nomarski prism and is shown in the schematic diagram, Fig. 6.26.

The recombination, i.e. removal of shear that takes place at the interference plane of the objective prism is of three types. The o–e-pairs passing through regions like 'A' where there is no phase object would reach the interference plane without any phase difference and therefore would be superimposed to give rise to a linearly polarised beam. The vibrational direction of it would be at 45° to the optic axis of the analyser and would cancel out. Pairs of o- and e-rays passing through 'B'

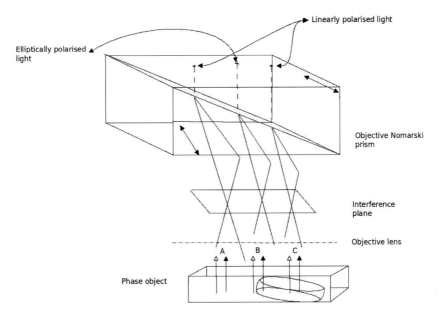

Linearly polarised light

Elliptically polarised light

Objective Nomarski prism

Interference plane

Objective lens

A   B   C

Phase object

Fig. 6.26 Objective Wollaston prism

regions would experience a gradient at the boundary corresponding to the phase object and the non-phase object as one of them would be passing from outside and the other from within. The resulting difference in optical path lengths would lead to an elliptically polarised beam when recombined. The analyser would admit part of the light (vectorial component that is parallel to its optic axis) and linearly polarise it. The amplitude that reaches the image plane will be picked up by the ocular and form an image. In region 'C' the ray pair completely passes through the phase object and any variation in optical properties or thickness of the phase object will affect both rays equally. Hence, the ray pair reaching the objective prism will be similar to that of 'A' and would be extinguished by the analyser. The optic axis of the analyser is oriented perpendicular, from front to back, to the orientation of the polariser. It would transmit linearly polarised rays that are parallel to its optic axis to the image plane located at the aperture plane of a tube lens or an ocular depending on the design. The ocular forms the final magnified image. Note that the above method of differential interference contrast does not improve or enhance the limit of resolution of the basic microscope.

Contrast mechanism: Optimum contrast is obtained by choosing a suitable shear distance. For illustration of the contrast mechanism, let's consider a hypothetical phase object with a uniform refractive index that induces optical path length differences of the type shown in Fig. 6.27. The optical path lengths induced by the specimen are plotted against 'x', the cross-sectional distance in (a). Then the contrast generated in the image would be negligible and regions r_1 and r_2 would appear with uniform contrast while r_3 would show some dip in contrast.

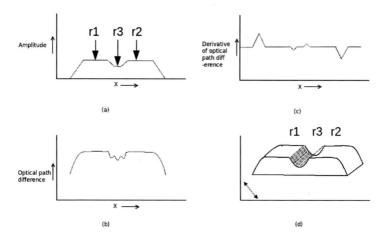

Fig. 6.27 Mechanism of virtual 3-D contrast in DIC images

If, on the other hand, a derivative of this curve is computed it would be as shown in (b). This differential interference contrast is achieved with the aid of closely spaced o- and e-pairs of rays whose velocities are different in the beam-splitter prism. The amplitude transferred to the interference plane would be as shown in (c) and the contrast in the image would be as in (d).

Retardation: We can enhance the contrast and also bring in a virtual 3-D effect by deliberately introducing a bias. For this purpose, a push-screw is provided at the objective Nomarski prism so that the prism can laterally be displaced with respect to the optic axis of the microscope either in -ve or +ve direction. The shear distance remaining the same, the phase difference curves corresponding to o- and e-rays get shifted with respect to each other as per the induced bias. As a result, the dark central fringe contrast will be replaced by a more uniform spread of grey contrast causing light and shadow effect. It is this effect that is responsible for the 3-D relief seen as elevations and depressions that are not truly present in the specimen. Of course, you need to know a priori whether such a structure is expected in the specimen. Notwithstanding this virtual effect, such contrast renders much clarity in the image. For example, the deformation twins occurring in Co-based superalloy can be clearly visualised in a DIC contrast image given in Fig. 6.28b as compared to the usual image obtained by normal optical microscopy as shown in (a).

Applications: The technique found for itself, a variety of metallurgical and materials applications such as for inspection of surface finish in critical applications. Figure 6.29 reveals a well-aligned set of dendrites in a TIG weldment of a cobalt-based superalloy obtained through DIC imaging.

It is a major tool for biological studies. Recently, developments have taken place in designing DIC microscope using infrared light.

Fig. 6.28 a Micrograph of
Co-based alloy MP35N
observed in conventional
bright field optical
microscopy after etching. **b**
Micrograph taken in a DIC
microscope shows the
deformation twins in clear
contrast. [Images are
provided through the
courtesy of Dr. S. V. S.
Narayana Murty, Materials
Characterisation Division,
Materials and Metallurgy
Group, Vikram Sarabhai
Space Centre,
Trivandrum-695 022, India]

(a)

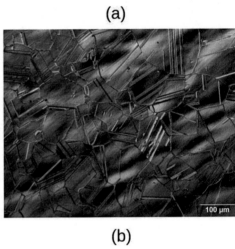

(b)

Exercises

Q1. A bulk isotropic crystal specimen, which had a polished flat surface initially,
requires to be examined in as-deformed condition under a microscope for any sig-
natures of deformation. Do you recommend (i) a polarised light microscope or (ii)
a differential interference contrast microscope or an interference-fringe microscope
for identifying the slip steps and for making an estimate of the step height? Give
reasons for your choice and explain the method.

Q2. (a) Explain the role of a field limiting aperture in deciding the limit of resolution
in an optical microscope. (b) Two convex lenses having focal lengths $f_1 = 20$ mm
and $f_2 = 40$ mm stand 90 mm apart in a compound microscope. If a tiny object
is placed 30 mm in front of the first lens, calculate the position of final image and
the total magnification. Illustrate your calculation in a neat sketch of the compound
microscope.

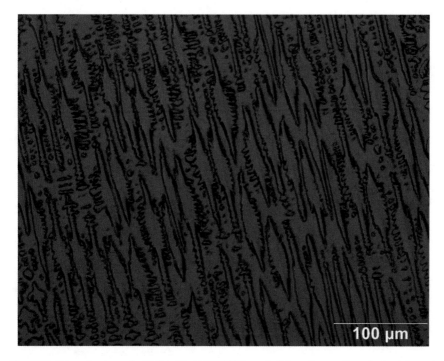

Fig. 6.29 Microstructure of a TIG weld of KC20WN, a cobalt-based superalloy obtained through DIC imaging in an Olympus GX71 inverted optical metallurgical microscope after etching. [The microstructure is provided through the courtesy of Dr. S. V. S. Narayana Murty, Materials Characterisation Division, Materials and Metallurgy Group, Vikram Sarabhai Space Centre, Trivandrum-695 022, India]

Q3. Explain with a neat sketch, why it is important to know the resolving power of an objective lens and not that of the ocular.

Q4. What is the scale on which you can get contrast in an optical image by using phase differences of the incident and reflected light rays?

Q5. If the step height of slip traces on the surface of a well-polished and subsequently deformed sample is of 1 μm on an average, what is the wavelength of light that you have to choose in an interferometer?

Chapter 7
Transmission Electron Microscopy

In this chapter, we will explore the contrast arising from various features of a specimen using the kinematical theory of diffraction contrast, and the necessary expressions will also be derived since the contrast we observe is counter-intuitive.The dynamical theory is not dealt with here as it falls outside the scope of this book. The features of interest are single crystals, i.e. single grains of a polycrystalline specimen, crystal defects of different dimensions and contrast from epitaxial layers or multilayered materials. Phase contrast, which arises in high-resolution electron microscopy is another important contrast mechanism and we will deal with it in some detail. In later sections, we will discuss, in brief, atomic number, differential energy and holographic and magnetic moment contrasts observable in TEM.

7.1 Importance of Contrast

Contrast is the difference in intensity of two features in an image, due to which we are able to distinctly recognise the presence of these two features of the image, as they are just resolved. If the features happen to have the same intensity, then it is impossible to distinguish them as two separate objects even if resolved. In transmission electron microscopy, it is often difficult to recognise even simple features such as grains, grain boundaries and dislocations merely by their contrast as the contrast itself varies with orientation (when the specimen is tilted in the goniometer). This dependence on orientation essentially arises due to the diffraction of the incident beam by the particular feature. Hence, this type of contrast is known as diffraction contrast. For example, a grain in a polycrystalline specimen may become completely featureless and bright when it is not in any diffracting orientation (i.e. not satisfying the Bragg angle for

© The Author(s), under exclusive license to Springer Nature Singapore Pte Ltd. 2022
G. V. S. Sastry, *Microstructural Characterisation Techniques*, Indian Institute
of Metals Series, https://doi.org/10.1007/978-981-19-3509-1_7

any crystallographic plane). The same grain when tilted to a diffracting orientation would appear dark in a bright field image. Stacking faults, special boundaries, dislocations, coherent precipitates and features similar to them show additional features of contrast that can only be understood by analytically calculating their contribution to the diffracted intensity using kinematical or dynamical (when a much rigorous interpretation is required) theory of diffraction. Mass-thickness contrast is minimal in the case of transmission electron microscopy.

7.2 Diffraction Contrast

We have derived an expression for intensity of a beam diffracted by a thin slab of crystal in Chap. 5 on Electron Diffraction as

$$I_o - I_t = I_s = |\phi_s|^2 = \left(\frac{\pi}{\xi_g}\right)^2 \frac{Sin^2(\pi ts)}{(\pi s)^2} \qquad (7.1)$$

where I_o is the intensity of the incident beam, I_t is that of the transmitted beam and I_s is that of the scattered beam. Here, only two beams, i.e. transmitted beam and only one diffracted beam are assumed to be present. Further, ξ_g is the extinction distance for that particular diffracted beam of the specimen material, 't' the thickness of the specimen at the location being sampled and 's' is the deviation from the exact Bragg condition for that plane in reciprocal space. In kinematical theory, it is assumed that diffraction is taking place slightly away from the exact Bragg condition (s) in addition to the assumption that $I_s << I_t$. These assumptions are satisfied to a large extent when thickness of the sample is very much less than ξ_g, i.e. of the order of ten nanometers. From Eq. (7.1), we observe that the diffracted intensity is dependent on two parameters, viz., thickness 't' of the specimen and the extent of deviation from the Bragg condition, 's'.

Those regions or features of the specimen that contribute significantly to the intensity of the diffracted beam would appear dark in the bright field (BF) image (and consequently would appear bright in the dark field (DF) image). That is the reason for naming this type of contrast as diffraction contrast. Although 't' is mentioned as one of the two variables on which the diffracted intensity depends, it should be realised that for a flat slab of crystal in a fixed orientation, neither 't' nor 's' are variables. Hence, the diffraction contrast arising in such a specimen is of uniform intensity (if the flat slab is devoid of any defects and is of a single-phase material). The sinusoidal variation in intensity expected as per Eq. (7.1) can only be realised in special geometries of the specimen cross-section and would be uniform otherwise as per the intersection of the bottom surface of the flat slab with the intensity plot (see Fig. 7.1).

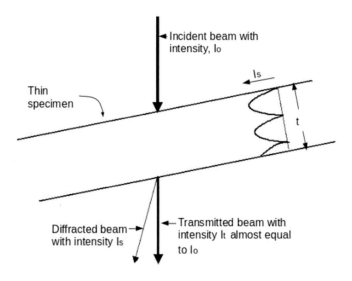

Fig. 7.1 Plot of Is verses 't'

7.2.1 Thickness Fringes

When the specimen assumes a wedge shape, as often seen at the edge of an electrolytically thinned sample, the incident beam experiences a variable thickness in its
path and so does the diffracted beam. Hence, at regular intervals of distance from
the edge, the diffracted beam becomes extinct in tune with the extinction distance
ξ_g of the material of the specimen and for that orientation of the specimen (**g**). This
happens till the thickness of the specimen reaches a uniform value within the width
of the transparent region. Equally spaced bright and dark bands of intensity will be
observed from the edge of specimen in the BF image. A typical image is shown in
Fig. 7.2.

The total number of bands multiplied by the extinction distance approximately
equals the thickness of the specimen.

$$n \times \xi_g = t$$

Fringe spacing and shape may vary from idealised parallel lines of uniform thickness according to the profile of the wedge sample and its thickness variations.

Fig. 7.2 Thickness fringes
in a wedge-shaped specimen

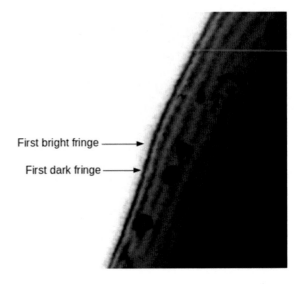

First bright fringe ⟶

First dark fringe ⟶

7.2.2 Bend Contours

The effect of varying 's' on the diffracted intensity can only be realised in a curved specimen having a uniform thickness as depicted in the schematic diagram(a) of Fig. 7.3.

The specimen is assumed to be in the exact Bragg condition for a (hkl) plane at the central location of the dome. Therefore, at the sloping surface on either side, the deviation from the exact Bragg condition is either positive or negative. According to Eq. 7.1, the intensity of the diffracted beam shows sinusoidal variations. As a result of these variations, dark and bright bands form a contour in the BF image as shown in Fig. 7.3b. These dark bands are called 'bend contours'. When such bending of the thin specimen is geometrically symmetric, it forms a hemispherical dome, and the corresponding bend contours intersect each other at the central point leading to the formation of a bend centre. The same is indicated in the boxed region of Fig. 7.3b. Bend contours and bend centres give valuable information regarding the local crystallographic orientation of the specimen in the BF image itself without the need to switch to the diffraction mode. For example, in Fig. 7.3b, the boxed region is in a perfect zone axis orientation that exhibits a two-fold rotation symmetry. We can also assess the stiffness of the material of the specimen by noticing the presence or absence of bend contours in the image. The absence of bend contours means the material is stiff since specimens of perfectly stiff materials such as oxides do not bend and cannot hence their specimens bend and would not show any bend contours in the image. Ramachandrarao and Sastry (Ramachanrarao and Sastry 1985) inferred the brittle nature of quasicrystals from the absence of bend contours in the BF images of the $Mg_{32}(Al,Zn)_{49}$ quasicrystalline specimens (see Fig. 7.4).

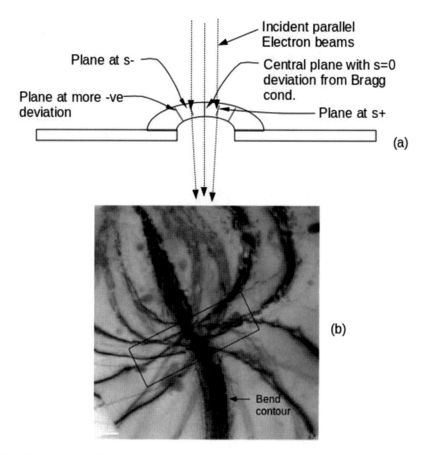

Fig. 7.3 **a** Schematic diagram of conditions of specimen leading to the formation of bend contours and **b** BF image of a bend centre

Fig. 7.4 Note the absence of bend contours in a quasicrystalline specimen (after P. Ramachandrarao and G. V. S. Sastry, Pramana J. of Phys. 25(1985), L225–L230. With permission)

The bend centres are also called zone axis patterns (ZAPs). ZAPs in reciprocal space, i.e. in diffraction patterns also show 2-f symmetry due to the reason that every $+\mathbf{g}$ diffracting vector has a $-\mathbf{g}$ due to the operative Friedel law. We also notice that fringes may be more widely spaced as a result of variation in $\mathbf{s}$ value, i.e. the bend becoming more gradual. Even if the curvature remains the same, we observe this difference in fringe spacing from one ZAP to another for the same (hkl) plane as we tilt the specimen from a low index ZAP to a high index one, the zones being widely spaced.

You may also notice that the bend contours are much more finely spaced lines compared to thickness fringes from the same specimen. You can understand this difference when you consider the relative influences of variation in 't' and in 's'. More complex contrast arises in actual images where interaction between bend contours and thickness fringes occurs. You need to invoke dynamical theory to understand this effect, where $s_{eff} = \sqrt{(s^2 + \frac{1}{\xi_g^2})}$.

7.2.3 Contrast from Crystals with Imperfections

The expression given in Eq. 7.1 is defined for a perfect crystal. But often materials contain defects, some of which are even inherent to their processing history or to their structure. Additional phase factor contributed by these defects need to be computed to understand the contrast arising out of these defects. Let $\mathbf{R}$ be the vector associated with the shift in atomic positions of the defect region. In the column approximation used in deriving Eq. 5.17, the crystal is perfect, up to the depth where the defect is being considered at present, and positional vector of the scatterer is represented by $\mathbf{r}_g$. At the location where the scattered beam enters the defected region of the lattice, the atomic locations need to be represented by a vector $\mathbf{r}' = \mathbf{r}_g + \mathbf{R}$. Then the phase factor assumes the form,

$$\mathbf{k} \cdot \mathbf{r}' = \exp((\mathbf{g} + \mathbf{s}) \cdot (\mathbf{r}_g + \mathbf{R}))$$
$$= \exp(\mathbf{g} \cdot \mathbf{r}_g + \mathbf{s} \cdot \mathbf{r}_g + \mathbf{g} \cdot \mathbf{R} + \mathbf{s} \cdot \mathbf{R})$$

In the above expression, $\mathbf{g} \cdot \mathbf{r}_g$ is always an integer and $\mathbf{s} \cdot \mathbf{R}$ is a negligibly small quantity.

Therefore,

$$A_s = \sum f_e \exp(2\pi i \mathbf{g} \cdot \mathbf{R}) \exp(2\pi i \mathbf{s} \cdot \mathbf{r}_g)$$

The phase factor can then be represented as

$$\Sigma F_e \exp(i\phi_{perfect}) \exp(i\phi_{defect})$$

$$i.e. \ \phi_{defect} = \exp(2\pi (\mathbf{g} \cdot \mathbf{R})) \tag{7.2}$$

Fig. 7.5 Stacking fault

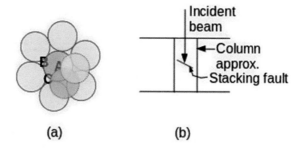

(a)   (b)

is the additional phase factor contributed by the defect.

Crystal defects in materials can be categorised as follows:

3-D: Volume defects such as stacking fault tetrahedra or coherent precipitates.

2-D: Stacking faults, twins, antiphase domain boundaries and grain boundaries in a poly crystalline material.

1-D: Dislocations.

0-D: Point defects such as vacancies and interstitials.

These are planar defects in the sequence of layers to be stacked to build a particular crystal. Stacking faults occur in abundance in certain f.c.c. and h.c.p. metals owing to their low stacking fault energy. An f.c.c. structure can be viewed as a stacking of ...ABCABC... close packed atomic layers normal to the [111] direction. Atomic stacking sites of a close packed layer can be marked as A, B and C, as shown in the diagram, Fig. 7.5.

The next close packed layer can be placed on top of the first one, with its A atom sitting in the trough formed at B and the third layer with its A atoms at site C. Thus, stacking continues as ...ABCABC... along the [111] direction if any of the layers is either missing or an extra layer occurs which is pushed in out of sequence, a stacking defect occurs. If the layer is missing, the fault is termed as intrinsic stacking fault, and if an extra layer is present, it is called extrinsic stacking fault. The shift vector associated with this fault is 1/6[11$\bar{2}$] or 1/3[111]. It can be atomistically thin unless another one occurs immediately below it in the thickness of the specimen. It is possible to obtain an expression for the intensity of scattered wave using the analytical equations or matrix method. We limit our discussions to the conclusions that can be drawn from Eq. 7.2.

From a qualitative interpretation, we can make a guess regarding the contrast to be observed in the BF image. The defect is automatically planar as discussed above and therefore divides the upper and lower portions of the crystal like a wedge surface, Fig. 7.5b. Therefore, we can expect alternate dark and bright fringes in the BF image just beneath the projected location of the defect in the specimen. The fringes are parallel bands bound on either side by curly boundaries as shown in Fig. 7.6 which is obtained from an austenitic stainless steel specimen (Courtesy: S. V. S. Narayana Murty, ISRO, Trivandrum). A detailed examination of the contrast features gives us additional information as listed here:

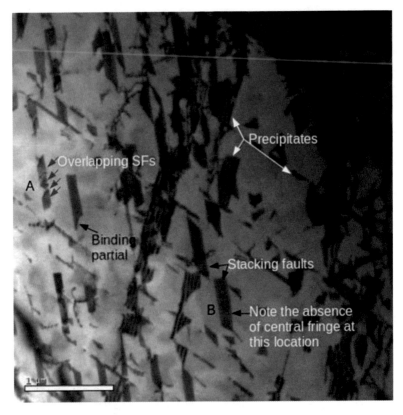

Fig. 7.6 BF image of stacking faults in an austenitic stainless steel (Figure provided through the courtesy of S. V. S. Narayana Murty, ISRO, Trivandrum, India. With permission)

1. The outer fringes are usually thicker and the central ones less in contrast if the stacking fault is in a thicker region.
2. The outer fringes may not always be dark in a BF image and the contrast depends on the sign of $\mathbf{g} \cdot \mathbf{R}$.
3. In the case of a fault that is lying inclined in the thickness of the specimen, the fringe corresponding to the top surface is either bright or dark depending on the sign of $\mathbf{g} \cdot \mathbf{R}$. The contrast of the top fringe remains the same in the DF image; on the contrary, the bottom fringe shows complimentary contrast.

From a systematic recording and examination of the observed contrast, it is possible to distinguish between extrinsic stacking faults and intrinsic ones (Carter and Williams (2009)). The fine details of contrast can only be interpreted by employing dynamical theory of diffraction. The stacking faults that we are considering in Fig. 7.6, in an austenitic stainless steel, may become invisible in certain orientation of the specimen in the microscope. From the expression ϕ_d (Eq. 7.2), we have

Fig. 7.7 Schematic diagram
explaining different contrast
features of stacking faults
shown in Fig. 7.6

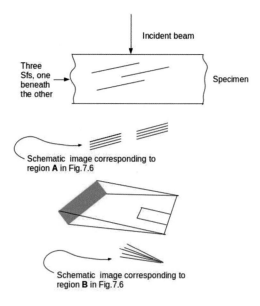

$$\phi_d = 2\pi (h + k + l) \cdot (u + v + w)$$

$$\text{i.e. } \phi_d = 2\pi \mathbf{g} \cdot \left(\frac{1}{6}[11\bar{2}] \right)$$

$$= \frac{\pi}{3}(h + k - 2l)$$

Therefore, if $(h+k-2l) = 6$ or $h+k = -2l$, the stacking fault will not contribute any additional phase factor and becomes invisible for all such orientations of the specimen (i.e. $\mathbf{g}$ and not the orientation of the fault in the specimen). It may also become invisible either in full or in part of it if three such stacking faults are present one below the other through the thickness of the specimen (Corresponding to region A in Fig. 7.6).

Three of them together will add an integral phase factor, leading to the effective absence of any additional phase factor, as shown in Fig. 7.7a. If the stacking fault happens to be lying inclined through the thickness of the specimen (Corresponding to region B in Fig. 7.6), as shown in Fig. 7.7b, pairs of fringes collapse in the image from inside to outside and terminate at a point on the edge of the thin specimen (if the fault is extending into the edge).

Antiphase domain boundaries (APBs): APBs are boundaries that separate domains or regions of one scheme of decoration of atoms on a superlattice cell from the other possible decorations. NiAl, CuAu and Cu_3Au are classic examples of these. If we consider the intermetallic, Ni_3Al, the ordered phase has Al and Ni atoms occupying corner and face centre positions, respectively, in one region and switching their positions in the adjoining region across an antiphase boundary. Here, the structure apparently looks like f.c.c., but is a superlattice based on a simple cubic lattice.

Otherwise, the stoichiometry should change to Al_3Ni. The shift vector associated with the thin planar defect is $R = 1/2[110]$ (see Fig. 7.8a). Therefore, when we image using g_{100}, we see fringe contrast in the image. Consider the ϕ_d expression once again,

$$\phi_d = 2\pi(h + k + l) \cdot \frac{1}{2}\left(1\hat{i} + 1\hat{j} + 0\hat{k}\right)$$

where $\hat{i}$, $\hat{j}$ and $\hat{k}$ are unit vectors.

$\phi_d = \pi(h+k)$ as h, k and l are 1, 0 and 0, the defect contributes a phase factor of π. Therefore, these boundaries are known as π boundaries.

APBs of saw-tooth shape are found in the case of Rutile unlike the ribbon-type boundaries seen in the case of Ni_3Al. The saw-tooth type is also observed in the case of rapidly solidified Na-Ag-Ge alloy (Fig. 7.8b) at A in the figure.

Twin boundaries: Mirror twins again have atomistically sharp boundaries similar to stacking faults. The twin plane (110) in f.c.c. crystal acts as a mirror across which the lattice on one side gets reflected on the other side of the twin region, i.e. lattice portions on either side of the twin are mirror related. Therefore, when the twin plane is parallel to the incident beam, the BF image shows a sharply defined boundary between the two twinned regions of a grain. Figure 7.9 shows a twin in such an orientation, in a Ni-base superalloy. If the twin boundary happens to be inclined to the incident electron beam, it exhibits typical wedge fringes (Contrast of a planar defect) as the twin region is rotated well away from the Bragg condition of the matrix (or twinned lattice on the other side of the mirror).

Nano twins: Existing mirror twins get deformed and new twins which are called deformation twins or mechanical twins that result from heavy deformation also occur in the f.c.c. metal silver. Figure 7.10a illustrates this effect in an as-deformed specimen (thin foil), parts of which are already so thin to be electron-transparent as such. One can notice the deformation structure within the twin plates. There are many other features observed in these specimens, which are discussed in a separate publication. (Foils of pure silver, called 'Panni', are produced in Varanasi by an age-old technique of hammering. Thin coins of silver are kept sandwiched between leather sheets and hammered with wooden mallets. The coin may occasionally be warmed-up in between to regain ductility, though not a necessary step always. Thanks are due to Sri Sandeep Nagar who provided these foils.)

Nano twins are also known to form under conditions of heavy deformation such as cryorolling in austenitic stainless steel UNS S31000. These are shown in Fig. 7.10b. Note that the deformation structure can be seen both in the matrix and twins. The twin width decreases significantly with the increasing percent reduction (Sarath kumar et al. 2020).

Transformation twins: In certain materials, a high-temperature, high-symmetry phase transforms into a low-temperature phase by symmetry breaking. The missing symmetry elements usually manifest as twins in the low-temperature phase. The high Tc superconducting compound $YBa_2Cu_3O_{7-\delta}$ is a typical example of it, where the high-temperature phase is tetragonal in its structure and transforms into an orthorhombic

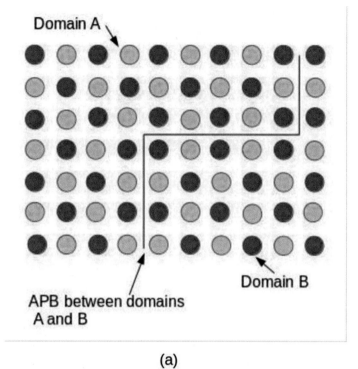

(a)

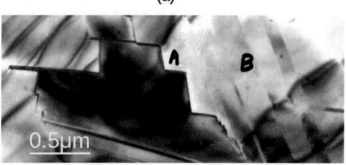

(b)

Fig. 7.8 a APBs in NiAl L12 and **b** Saw-tooth-type APBs, as at A in rapidly solidified Na-Ag-Ge (after Pasupathy Shankar, 1984, B.Tech. Project, IIT (BHU), Varanasi, India. With permission)

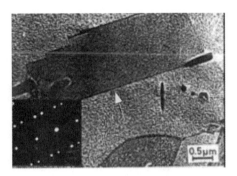

Fig. 7.9 Twins in Ni,Fe-base super alloy (after A. Bhattacharyya, G. V. S. Sastry and V. V. Kutumbarao, Journal of materials science 34(3), (1999) 587–591. With permission)

superconducting phase at low temperature with a slight adjustment of the basal plane. The a and b lattice parameters of the basal plane of the orthorhombic phase are of almost similar value. This symmetry breaking leads to the formation of twins in the a-b plane. Such twins, as shown in Fig. 7.11, are known as structural twins or transformation twins. They appear as dark and bright bands when imaged down the c-axis [001] direction.

Grain boundaries: Grain boundaries in a polycrystalline material are also 2-D defects. They are planar but curved in the 3-D. Thus, they lack a fixed shift vector **R**. Nevertheless, as two adjoining grains with a high angle grain boundary are never in the same crystallographic orientation, a wedge effect is present and gives rise to fringe contrast for the boundary as shown in Fig. 7.12.

Line defects: They are most extensively studied defects by electron microscopy because they play a greater role in deciding material behaviour. A line defect can be assigned a line vector **l**, a unit vector along the line, which may be a straight line in a simple model case or a straight line segment of a curved one. In its pure form, the line defect is of two types, (i) an edge dislocation and (ii) a screw dislocation. At the core of a dislocation, atomic displacements associated with the defect are maximum though the strain field extends beyond the core. The defect vector **R** associated with the atomic displacement of the line defect is given by Hirth and Lothe (1982).

$$\mathbf{R}(r, \theta) = \frac{1}{2\pi} \left(\mathbf{b}\theta + \frac{1}{4(1-\gamma)} \{ \mathbf{b}_e + \mathbf{b} \times \mathbf{l}(2(1-2\nu)ln\gamma + Cos2\theta) \} \right) \quad (7.3)$$

in polar coordinates as shown in Fig. 7.13, where **b** is the Burgers vector, $\mathbf{b}_e$ is the edge component of the Burgers vector, **l** is the unit vector along the dislocation line associated with the defect and ν is Poisson's ratio. Let us first consider the defect in its pure form. We can now define the pure edge and pure screw dislocations by taking a crystal block subjected to (i) deformation under shear, Fig. 7.14a, and (ii) by wedging an extra half-plane of atoms in between a set of parallel planes of a crystal, Fig. 7.14b.

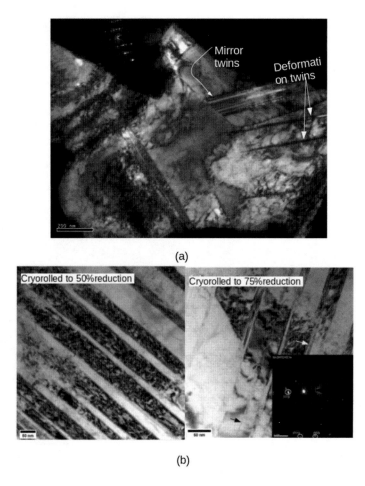

(a)

(b)

Fig. 7.10 a Nano twins in heavily deformed silver. **b** Nano twins in cryorolled austenitic stainless steel. Insets show corresponding diffraction patterns. (after G. Venkata Sarath Kumar, K. R. Mangipudi, G. V. S. Sastry, Lalit Kumar Singh, S. Dhanasekaran and K. Sivaprasad, Nature Scientific Reports, 10:354 (2020). With permission)

Fig. 7.11 Transformation twins in $YBa_2Cu_3O_{7-d}$ (after G. V. S. Sastry, R. Wördenweber and H. C. Freyhardt, Journal of Applied Physics 65, (1989) 3975. With permission)

Fig. 7.12 Fringe contrast at grain boundaries

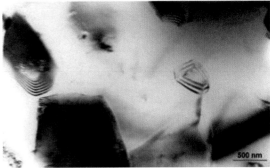

Fig. 7.13 Displacement of atomic planes in the neighbourhood of a dislocation

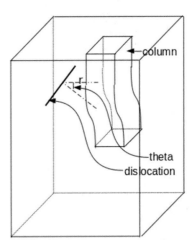

Screw dislocation: In Fig. 7.14a, due to the shear stress, a portion of the crystal is sheared from left to right creating a screw dislocation (atomic displacements around the core forming a helix) at AA. The screw dislocation moves in its glide plane, perpendicular to the line vector **l** and leaves a deformation step parallel to itself (or

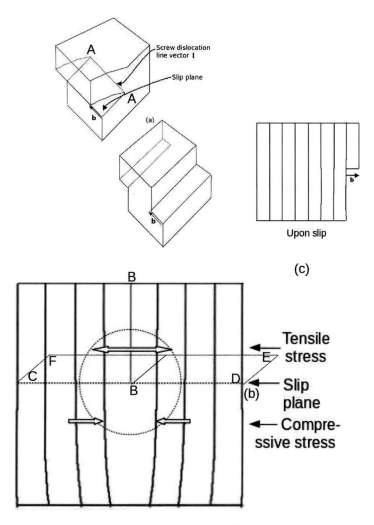

Fig. 7.14 schematic model of **a** a screw dislocation and its slip-step in the crystal equal to its Burgers vector and **b** of an edge dislocation and its slip-step in the crystal equal to its Burgers vector

perpendicular to its direction of motion.). Thus, the Burgers vector is parallel to the line vector **l**. Therefore, $l \cdot b = |b|$.

Edge dislocation: It can be realised from Fig. 7.14b above that the wedged-in extra half-plane creates tensile stress on the adjacent planes parallel to it in the upper half of the crystal, i.e. above the slip plane and a compressive stress on the same planes in the lower half of the crystal. In fact, there is a disturbance of atomic arrangements in the planes perpendicular to the extra half-plane which cannot be realised in the diagram. Therefore, the edge of the extra half-plane BB is defined as the edge dislocation.

This cannot be a full plane, since in such a case it would increment the 'thickness' by one 'd' spacing. When the extra half-plane is pushed either to the left or right of the crystal by the application of a shear force along the slip plane, it leaves one atomic step at that end of the crystal. This single atomic step is defined as the Burgers vector $\mathbf{b}$ of the dislocation. Therefore, $\mathbf{b} \cdot \mathbf{l} = 0$ for the edge dislocation, and it glides on the slip plane leaving a step equal to $|\mathbf{b}|$ in the same direction. From the above definitions, it is clear that $\mathbf{R}$ in Eq. 7.3 takes the form

$$\mathbf{R} = \frac{1}{2\pi}(\mathbf{b}\theta)$$

Since $\mathbf{b}_e = 0$ and $\mathbf{b} \times \mathbf{l} = 0$

$$i.e. \ \mathbf{R} = \frac{1}{2\pi}\left(\mathbf{b}\mathrm{Tan}\left(\frac{z-z'}{x}\right)\right)$$

For a pure edge dislocation, $\mathbf{b} = \mathbf{b}_e$ and the $\mathbf{b} \times \mathbf{l}$ component always survives. *Invisibility criterion*: Now let's take help of Eq. 7.2 to interpret the contrast arising from a screw dislocation. The phase factor contributed by the defect would be

$$\phi_d = \mathbf{g} \cdot \mathbf{R} \propto \mathbf{g} \cdot \mathbf{b}$$

Depending upon the operating reflection (remember, we are keeping the specimen in two-beam condition), the dot product $\mathbf{g} \cdot \mathbf{b}$ takes integer or fractional values. For an isotropic crystal and for all non-zero values of 'n', the dislocation shows contrast, and for zero value, it becomes *invisible*. In this condition of invisibility, the dislocation lies in the diffraction plane in the real space with any arbitrary orientation. From the model shown in Fig. 7.14a, it is also evident that the plane containing the screw dislocation (the line vector $\mathbf{l}$ of it) is not having any component of $\mathbf{R}$. If we identify two $\mathbf{g}$ vectors which satisfy the invisibility criterion by tilting the specimen to two-beam conditions, we can obtain the Burgers vector $\mathbf{b}$ of the screw dislocation upon taking the cross product of those $\mathbf{g}_1$ and $\mathbf{g}_2$. Invisibility criterion for an edge dislocation is a more involved procedure. The dislocation would still be visible even if $\mathbf{g} \cdot \mathbf{b}_e = 0$ unless $\mathbf{g} \cdot \mathbf{b}_e \times \mathbf{l}$ is also equal to zero. This additional requirement arises since not only the planes parallel to the wedge-plane but also the planes perpendicular to it are distorted. The 111 family of planes constitute the slip planes in f.c.c. crystals. If the wedge-plane BB defines the edge dislocation by the bottom edge of it, the line vector $\mathbf{l}$ and the Burgers vector $\mathbf{b}$ are located as shown in Fig. 7.14b. A vector $\mathbf{n}$ which is normal to the appropriate 111 plane is defined by $\mathbf{b} \times \mathbf{l}$. Therefore, we can find all those poles, i.e. all those $\mathbf{g}$ vectors that lie normal to $\mathbf{n}$ by taking help of the standard projection of a cubic crystal along [111]. They project on to the equatorial circle and are of the type $(\bar{1}10)$, $(1\bar{1}0)$ and $(11\bar{2})$ with appropriate signs. Out of these, we can find the $\mathbf{g}$ vectors that not only satisfy $\mathbf{g} \cdot \mathbf{b}_e = 0$ but also $\mathbf{g} \cdot \mathbf{b} \times \mathbf{l} = 0$. For all other $\mathbf{g}$ vectors which only satisfy $\mathbf{g} \cdot \mathbf{b}_e = 0$, there would be residual contrast in the image.

As far as contrast is concerned, the dislocation would normally appear dark in a bright background of the BF image. We should remember that while ϕ_{defect} only tells us about the phase factor contributed additionally by a dislocation, there are other variables in the perfect crystal phase factor, $\phi_{perfect}$ such as $\mathbf{s}$, even when the thickness is constant. Thus, a bend contour in the region containing dislocations would show those dislocations that lie at its central region ($\mathbf{s} = 0$) in bright contrast and larger in width in a DF image as against the narrow and less brighter appearance of those that lie in the outer edges of the bend contour where $\mathbf{s}_g$ is either +ve. or −ve. In order to gain further understanding of the contrast effects, one needs to consider the depth dependence of $\mathbf{R}_z$ in the column approximation shown in Fig. 7.13. It means that the strain field created by the dislocation would influence the phase factor up to a distance 'x' in the crystal through the thickness 'z' of the crystal. The rate of change of defect phase factor with dz, a thickness element in the column approximation, also varies as a function of these two parameters in addition to its dependence on the effective deviation parameter, $\mathbf{s}_{eff}$. It takes the form

$$\frac{d\phi_g}{dz} = \frac{\pi i}{\xi_g}\phi_\circ + 2\pi i \mathbf{S}_R^{eff} \phi_g \qquad (7.4)$$

where $\phi_\circ$ is the transmitted wave phase factor, ϕ_g is the phase factor of scattered wave and $\mathbf{S}_R^{eff} \phi_g$ represents

$$\mathbf{S}_R^{eff} \phi_g = \mathbf{s} + \mathbf{g} \cdot \frac{d R}{dz}$$

These two formulae are very effective in explaining various finer details of contrast associated with linear defects. A discussion on the full implications of these is beyond the scope of the present text, and interested readers are referred to discussions in Williams and Carter (2009) and de Graef (2003). One important aspect regarding the location of image vis-a-vis actual position of the dislocation in the specimen will, however, be taken up here. When we consider the products ($\mathbf{g} \cdot \mathbf{b}$), we notice that depending on the sign of it, the DF image location shifts either to the right or left of the projected location of the dislocation in the image. The relative positions are shown in Fig. 7.15. The change of sign can be there due either to $\mathbf{g}$, $\mathbf{b}$, $\mathbf{s}$ or $\mathbf{R}$. Usually, we keep $\mathbf{s}$ positive while imaging a dislocation (Convention: $\mathbf{s}$ is taken as positive if the Ewald sphere lies outside the $\mathbf{g}$ vector). In the case of a non-isotropic crystal, this analysis doesn't hold good even for pure edge or pure screw dislocations and requires to be modelled. So is the case if the dislocation is of a mixed type or if dynamical conditions prevail.

Volume defects: At the nascent stage of precipitation in a solid state in an alloy, the precipitate atoms segregate on certain atomic planes of the matrix solid solution that possesses similar atomic packing as required for the precipitate crystal structure. The precipitate shares these matching atomic planes of the matrix while nucleating itself, though bending them slightly suits its own interplanar spacings. As a result of this bending, a strain develops in the matrix lattice. Usually, these precipitates have a

Fig. 7.15 Displacement of
image with respect to the
dislocation in the crystal

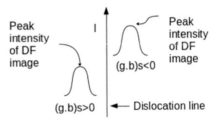

disc shape at this stage (e.g. G-P zones of Al-Cu system) though a spherical shape is
also observed (e.g. Al-Ag system). If the lattice parameters of the two phases differ
slightly, then the volume strain or lattice misfit δ is given by

$$\delta = \frac{a_{ppt} - a_{mat}}{a_{mat}}$$

where a_{ppt} is the lattice parameter of the precipitate phase and a_{mat} that of the matrix
for a cubic phase. If $a_{mat} > a_{ppt}$, then precipitate exerts tensile stress on the lattice
planes of the matrix. The shift vector $\mathbf{R}$ and associated contrast effects were worked
out by Ashby and Brown (1963) according to whom $\mathbf{R}$ for a spherical particle can
be expressed as

$$\mathbf{R(r)} = C_e \frac{r_o^3}{|\mathbf{r}|^3} \mathbf{r}$$

where C_e is a constant that expresses the combined effect of elastic constants, r_o is
the radius of the particle (assumed spherical) and $\mathbf{r}$ is the positional vector.

$$C_e = \frac{3k\delta(1 + v)}{3k(1 + v) + 2E}$$

where k is the bulk modulus, δ is misfit strain (assumed to be completely in Matrix!,
i.e. particle is hard), v is Poisson's ratio and E is elastic modulus. If the shape of
the particle is not spherical, the displacement field needs to be modelled separately.
Since $\mathbf{R}$ is radially symmetric, the contrast in the BF image would be the appearance
of dark lobes on either side of a central line-of-no-contrast in between them. Figure
7.16 shows such contrast in the case of a superalloy.

Since the entire strain is located in the matrix ,the precipitate region shows a line-
of-no-contrast, i.e. $\mathbf{g} \cdot \mathbf{R} = 0$. This type of contrast is also known as 'coffee-bean'
contrast or 'doughnut' contrast due to its appearance.

Weak-beam imaging of defects: This is a technique rather than a new theory of
diffraction contrast for weak diffracting conditions. It is mostly practised in DF mode
for imaging defects, particularly the line defects. You may refer to Sect. 4.3 in which
we discussed the formation of centred dark field image, for understanding the ray
diagrams discussed in the foregoing. In Fig. 7.17a, we see the positions of transmitted
and diffracted beams in a two-beam condition. For obtaining a weak-beam condition,
we tilt the transmitted beam to $-\mathbf{g}$ position using the DF tilt controls (NOT THE

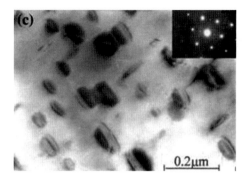

Fig. 7.16 Coffee-bean contrast of coherent precipitates in a Ni-base superalloy (after K. V. U. Praveen, G. V. S. Sastry and V. Singh, Trans. Indian Inst. Met., 57, No. 6, (2004) 623–630. With permission)

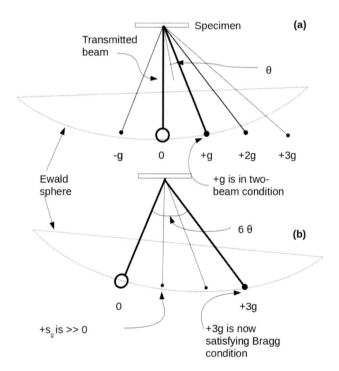

Fig. 7.17 Ray diagram for weak-beam dark field imaging, (a) initial two beam condition, (b) after tilting to **g**, 3**g** condition

SPECIMEN CONTROLS!) by an angle $+2\theta$. This tilts the incident beam by an angle of 3θ in the real space which is obvious from the diagram (b), so that the particular reflection **g** is experiencing a very large deviation from the exact Bragg condition, i.e. $\gg +s_g$.

Fig. 7.18 Weak-beam DF images of a-type dislocations in Ti$_2$AlNb orthorhombic phase (after T. K. Nandy and D. Banerjee, Intermetallics, 8(2000), 915. With permission)

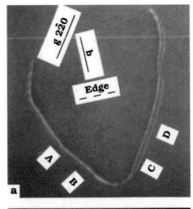

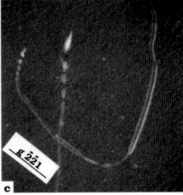

In turn, the image formed corresponds to that location where the derivative of **g** · **R** is very large (Eq. 7.4). Therefore, the image is called weak-beam dark field image (Fig. 7.18). Since the incident-beam angle is 3θ for the same set of planes, the beam that satisfies the Bragg angle with respect to the transmitted beam will be 6θ away, i.e. 3g beam. It is for this reason the technique is also called **g**-3**g** DF imaging. While the technique is extremely useful in quantifying the dislocation characteristics such as their spacings in an array, precise estimation of separation of dislocation partials and spacing of superdislocation pairs, it is also a difficult technique owing to the very long exposure times required for recording of the images. There are attendant signal-to-noise ratio problems (due to thermal shift of the sample, etc.). For any such quantification, **s** is kept at 0.2 nm^{-1}. For further details, readers are referred to Williams and Carter (2009).

7.2.4 Contrast from Epitaxial Layers

Multilayer materials have gained much importance in recent times for device applications like sensors. Each constituent of it may be only a few nanometers in thickness. Such layering can also occur due to the epitaxial growth of precipitates. A prerequisite of surface nucleation is the availability of suitably oriented atomic planes in the substrate material with matching 'd' values. If d_1 is the interplanar spacing in the epitaxy with $\mathbf{g}_1$ parallel to the surface and d_2 is the interplanar spacing in the substrate with $\mathbf{g}_2$ oriented parallel to $\mathbf{g}_1$, then the incident beam gets doubly diffracted giving rise to additional spots close to those of the substrate reflections including the transmitted beam. Thus, the beams collected by the objective aperture include the central transmitted beam and the additional beams around it. They interfere to give rise to an interference pattern in the BF image. The fringes formed by the interference are named as the Moiré fringes. They are fairly sharply defined compared to thickness fringes and can sometimes lead to confusion with dislocation arrays. The fringe spacing D is given by

$$D = \frac{d_1 d_2}{|d_1 - d_2|}$$

If the two $\mathbf{g}$ vectors are of the same magnitude, i.e. layer-wise deposition of the same material or two grains with a small angle of twist on top of each other, then we obtain a rotation Moiré in the image. Fringes occur perpendicular to $\delta\mathbf{g}$. The fringe spacing D (Fig. 7.19) is given by

$$D \approx \frac{d}{\beta}$$

where β is the angle of twist between them expressed in radians. Note that $\mathbf{g}_1$ and $\mathbf{g}_2$ are the same magnitude-wise and differ only in their orientation (i.e. diffraction spots occur with a small angular separation but with the same radial distance 'r' from the transmitted spot).

Fig. 7.19 Rotation Moiré in an aluminium alloy

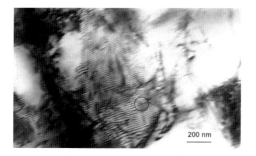

200 nm

A third type of Moiré occurs when there is both difference in magnitude of **g** vectors and their orientations exist. Under such conditions, the two layers are rotated with respect to each other and they give rise to mixed Moiré.

$$D_{mix} \approx \frac{d_1 d_2}{\sqrt{(d_2 - d_1)^2 + d_1 d_2 \beta^2}}$$

and the fringes will be in a direction $\perp$ to the $\delta\mathbf{g}$.

A magnified view of 'd' spacings is achieved in all three cases, and hence, any defects such as dislocations present in either of the two layers get projected with a magnified view, making Moiré patterns very useful for their study. With the advent of sample preparation techniques such as precision ion mill and focussed ion beam machines (FIB), it is now possible to prepare 3 mm disc samples of multilayers by making sandwiches of several specimens. Cross-sectional microscopy has eased many of the earlier-time difficulties in studying interface structures of multilayers.

7.2.5 Contrast from Polymeric Materials and Polymer-Based Composites

We have examined in some detail the diffraction patterns from amorphous phases of different materials in Sect. 5.8. Images of amorphous phases are usually contrastless, unless the amorphous phase happens to be actually microcrystalline with short range order than being truly amorphous (disordered with random atomic structure).

MWCNT and Graphene Nanofillers

Polymeric materials can be completely amorphous or crystalline depending on their processing conditions in which their long chains of molecules are either randomly oriented or are aligned in a direction. An illustration of it can be seen in (Fig. 7.20) the case of thermoplastic polyurethane, TPU, (Sharma et al. 2021), which consists of phase-separated, hard-segment domains dispersed in a soft matrix. The hard constituent is crystalline and an amorphous phase constitutes the soft matrix as evidenced by the electron diffraction pattern given in (D) of Fig. 7.20.

In the pattern, diffraction rings due to the crystalline phase are dominant over the diffuse rings arising from the amorphous phase. Light speckled contrast in the image (7.20a) is all the evidence that one can get about the phase-separated state of the polymer, even at high resolution. Whether this speckle corresponds to the phase-separated state of the hard-crystalline and soft components needs to be confirmed by additional evidence.

The alignment and length scale of the phase-separated components of a polymer can be enhanced by the addition of suitable nanofillers to enhance the performance

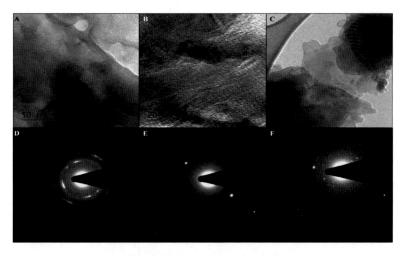

Fig. 7.20 TEM images of the cryomicrotome samples of pristine TPU (A), TPU reinforced with 0.5 wt and hydroxylated edge-functionalized few-layer graphene (C); the corresponding SAD patterns are illustrated in (DF), respectively. (after Kavita Sharma, Debi Garai, Ajay Gupta, Debmalya Roy and N. Eswara Prasad, J. Phys. Chem. C, 125(2021) 21653–21662. With permission)

of a polymer by the synergistic effect of a composite. MWCNTs or suitably treated multilayer graphene are selected to modify the microstructure of TPU (Sharma et al. 2021).

The effect of these additions can once again be assessed by transmission electron microscopy. Figure 7.20b of TPU reinforced with nanoscale MWCNTs shows more enhanced speckle as compared to the non-treated polymer and the corresponding diffraction pattern (Fig. 7.20e) indicating crystalline peaks replacing in place of the polycrystalline rings of the non-treated specimen. This is an indication of a high degree of alignment of the coherently diffracting MWCNT nanofiller. Diffuse rings of the matrix amorphous phase of the polymer are prominent in the pattern in the absence of the dominant polycrystalline rings. The authors, however, attributed the diffuse rings to the anisotropic distribution of the high-density hard segments of the TPU driven by the attachment to the nanofiller. Additionally, the microstructure shows short segments of MWCNTs that are apparently unconnected to the hard segments of TPU (Observe the speckle orientation).

The crystalline filler is a 2-D, hydroxylated, edge-functionalised, multilayer graphene in the second category of TPU-based composite studied by Kavita Sharma et al. (2021). As a consequence of 2-D nature of the filler, one can observe Moiré fringes at regions corresponding to the crystalline filler in the image (Fig. 7.20c). Diffraction from such typical regions, (Fig. 7.20f) shows many crystalline peaks superimposed over the diffuse rings from the amorphous phase.

Two more examples of composites can be taken up here, of which one is $FeCo_3$ nanoparticles dispersed in a cross-linked polydimethylsiloxane elastomer (Mordina

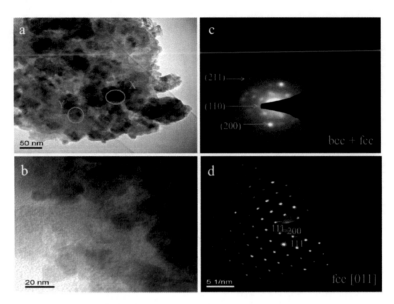

Fig. 7.21 a, b TEM images of FeCo₃ nanoparticles at two different magnifications; **c** selected area electron diffraction pattern of FeCo₃ nanoparticles taken over a large area showing both bcc and fcc reflections and **d** [011] electron diffraction pattern of FeCo₃ particles showing the fcc reflections in [011] zone axis. (after Bablu Mordina, Rajesh Kumar Tiwari, Dipak Kumar Setua and Ashutosh Sharma, J. Phys. Chem. C 118(2014) 25684–25703. With permission)

et al. 2014), and the other is epoxy-based microwave-absorbing composite containing $CoFe_2O_4$ nanoparticles (Jaiswal et al. 2020).

FeCo₃ Nanofiller

Magnetorheological elastomers have several applications, the most important one being a tunable vibration absorber in mechanical devices. These smart materials constitute a uniform dispersion of magnetic nanoparticles that form a 3-D network under the influence of a magnetic field. The performance of the device depends on the optimum quantity of elastomer and size distribution of the dispersed magnetic nanoparticles. An example of Polydimethylsiloxane + FeCo₃ composite (Mordina et al. 2014) is chosen to illustrate the application of transmission electron microscopy in assessing the microstructure and crystal structure of the constituents of the composite. These studies showed that FeCo₃ exists as a phase mixture of Fe-rich particles with a b.c.c. structure and Co-rich particles with an f.c.c. structure as shown in Fig. 7.21a.

CoFe₂O₄ Nanofiller

The third example of a polymer-based nanocomposite is an epoxy in which ternary core-shell nanofiller ($CoFe_2O_4$/rGO/SiO₂) is dispersed by microwave irradiation.

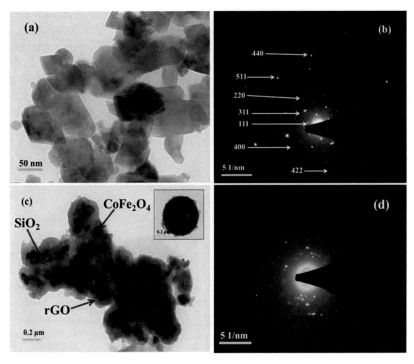

Fig. 7.22 TEM images and SAED patterns of **a, b** CoFe$_2$O$_4$ and **c, d** CoFe$_2$O$_4$/rGO/SiO$_2$ nanofiller. (after Rimpa Jaiswal, Kavita Agarwal, Vivek Pratap, Amit Soni, Subodh Kumar, Kingsuk Mukhopadhyay and N. Eswara Prasad, Materials Science & Engineering B262(2020) 114711. With permission)

Such composites are sought after for their application as lightweight microwave-absorbing material. The Cobalt ferrite is coated with reduced graphene oxide by microwave irradiation and finally by SiO$_2$.

Besides knowing the surface morphology and particle size distribution of the coated particles by a Field Emission Scanning Electron Microscope, it is also important to verify the crystal structure and through-thickness morphology of the nanofiller.

Transmission electron microscopy was employed for this purpose to establish that ferrite particles are almost cubic in their morphology and possess the typical spinel structure (see the diffraction pattern given in Fig. 7.22b). The coated nanoparticles (Approx 200 nm) show a clear three-layer structure (shown as an inset in Fig. 7.22). The diffraction pattern in (d) reveals the amorphous nature of SiO$_2$ along with crystalline peaks originating from the cobalt ferrite particles.

The key point in the study of polymeric materials by TEM is that one may not get much information from the image contrast but diffraction evidence can be gathered if there are crystalline components in the polymer or polymer-based composite. Another issue is related to the techniques to be used for extracting thin film specimen from bulk material. Usually, cryomicrotomy is adopted with a glass knife-edge or a diamond knife-edge. This technique involves chiselling out thin wafers by a knife-edge from a bulk stock kept at cryo temperatures to make it brittle. More sophisticated techniques such as Focussed Ion Beam (FIB) which employs Ga or other ion beams yield a highly precise cut from the bulk sample. Readers are suggested to refer to Handbooks on the topic of Sample preparation (e.g. Sample Preparation Handbook for Transmission Electron Microscopy: Methodology and Techniques (Vol. 1 & 2), Jeanne Ayache, Luc Beaunier, Jacqueline Boumendil, Gabrielle Ehret and Daniele Laub, Springer 2010, 2014).

7.3 Phase Contrast

When two electron beams travelling parallel through a specimen with a phase difference between them interfere, they form a pattern of parallel lines (lattice fringes, see Fig. 7.23) depending on the extent of the phase difference. This contrast is known as phase contrast.

Diffraction contrast also gives rise to alternate dark and bright fringe contrast under certain conditions of the specimen orientation (refer to discussions in previous Sect. 7.2) (e.g. thickness fringes, stacking fault fringes, etc.). In these examples also, the contrast is essentially due to interference of parallel beams travelling through the thickness of the specimen. The difference between the two types lies in the number of beams used for the image. You need to include at least two beams in the objective aperture for obtaining phase contrast in the image while for forming images showing fringe due to diffraction contrast you include one beam only. This may be either a transmitted beam, for obtaining BF image or diffracted beam for DF image. The resulting image (in BF) has bright regions in which certain features which are strongly diffracting appear dark, as the objective aperture stops them from reaching the image. Contrast arises due to attenuation of the amplitude of transmitted beam in that direction and hence is known as amplitude contrast. The contrast obtained in phase contrast images is weak and also comparatively difficult to interpret without image simulation. Besides these difficulties, the microscope needs to be in fully aligned and clean condition, as you would see in later sections, how optical misalignment and aberrations affect the resolution of the image. The theory developed for calculating the image contrast is based on information theory since the microscope is expected to function as a linear system of the information processor.

7.3.1 Requirements on the Part of Microscope

Details given in the following may appear trivial, descriptive and non-analytical factors for a beginner, but as you continue to practise high-resolution imaging, you would realise the importance of these.

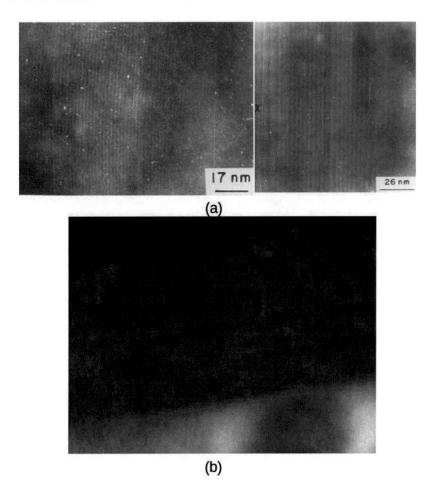

Fig. 7.23 a Lattice fringes of a rapidly solidified intermetallic phase in Al-Pd alloy. Note the variation of fringe spacing at X. **b** Lattice fringes (of graphite layers) in MWCNTs (Micrograph provided through the courtesy of Professor P. M. Ajayan. Work carried out by the author while at Rensselaer Polytechnic Institute, Troy, on B. H. Kulkarni Memorial Visiting Fellowship from Infosys Foundation)

Vacuum: Routine, low-resolution microscopy itself demands a clean high vacuum at the specimen. The requirements of high-resolution electron microscopy (HREM) are more stringent, and a vacuum better than 10^{-7} is preferred and that too oil-free. Often the specimen surface, acting as a catalyst, breaks the oil vapours coming from the diffusion pump to lower hydrocarbons and carbonaceous material. The products of dissociation deposit as a layer on the specimen surface. This deposit leads to a loss in image intensity while the area is being examined, resulting in longer exposure times for recording the image. That is why experienced microscopists work on an area similar to and adjoining the area of interest for setting up the requisite parameters such as the tilt of the specimen, contrast and correction of astigmatism and shift to the area of interest for recording the image. Anti-contamination devices are available on all microscopes, but the bubbling of LN_2 in the dewar transmits vibrations which interfere with image stability.

Specimen: Foremost requirement with respect to the specimen is its thickness which should not be greater than 10 nm even for a low atomic number. Its surface should be devoid of any contaminants coming from the sample thinning procedures adopted. These contaminants are either the reaction products of the electrolyte in case of electrolytic thinning or decomposed oil vapours in ion beam thinning which could easily be removed by plasma cleaning. The cleaned specimens should be mounted in the desired specimen holder and either loaded in the microscope or kept in a vacuum desiccator. Powder specimens should be mixed in a cold-setting resin and then ion milled. It should be ensured that the specimen is firmly mounted in the holder, completely dry and lint-free before loading it in the microscope.

Specimen holder: The 'O'rings of the specimen holder should not be excessively greased, and often the fat layer on the fingertips is sufficient to grease them. Before loading the holder, the tip of the holder may be plasma cleaned and the 'O'ring may be inspected for any lint sticking on it. Any such fibre or dirt on it would result in leakage of vacuum in addition to mechanical instability-driven drift of the holder. Thermal shocks, air turbulence created by air conditioners, acoustic noise (Since the microscope acts as a microphone. You can test by clapping your hands near the column and observe the image vibrating.) and vibrations transmitted by the ground all affect the image and should completely be avoided at least while recording the image. It is a good practice to switch off the air conditioners while recording the images if that doesn't change the ambient temperature too much. Most of these requirements are taken into account by the installation personnel while designing the layout of various ancillary equipment and the microscope column itself.

Objective lens: It is the primary lens that decides the achievable resolution in the phase contrast image. Hence, all the optical aberrations associated with a convex lens, as discussed by us in Chap. 6, are also present in the objective lens of the transmission electron microscope. Some of them can be eliminated by proper axial alignment of the microscope starting from gun to objective lens. Spherical aberration cannot be corrected (Since no concave 'magnetic lens' is available to combine with) except in the recently developed aberration-corrected microscopes. It is taken care of in an indirect way while formulating the equations for phase contrast which will be discussed in Sect. 7.3.2. Condenser lens astigmatism is corrected while aligning the microscope, but objective astigmatism requires to be corrected online at the region of the specimen being imaged. This can easily be done using the thin amorphous edge of an electrolytically thinned specimen by adjusting the objective stigmators and taking an FFT (Fast Fourier Transform, can now be obtained instantaneously—thanks to the computer interfacing, the microscope functions are made available in the recent models. It used to be a day-long effort with laser optical benches and photographic films earlier). The FFT shows amorphous rings that appear either elliptical or even more distorted if objective astigmatism is present and would become perfectly circular if the objective astigmatism gets fully corrected (see Fig. 7.24).

Fig. 7.24 **a** Stigmated rings of diffraction pattern from amorphous carbon (schematic) and **b** Actual pattern (FFT taken online) after correcting the same. Note the perfect circular appearance

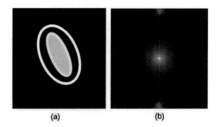

(a)　　　(b)

It is highly recommended that the objective lens is operated in its fully saturated condition. The specimen can be raised in the column by adjusting the z-coordinate (height) of the goniometer stage to bring the image to focus, rather than changing the objective lens current. From this point, the image contrast can be manipulated by changing the Δf, focal length through a change in lens current.

7.3.2 Image Formation

In a thin specimen (<10 nm, for low and medium atomic number elements), the incident electrons interact only with the projected potential in the forward direction, i.e. z-direction, which means the side-ways spread of the scattering process is almost non-existent. The projected potential, therefore, influences the phase change and can be expressed as a convolution of the projected potential with an interaction constant $\sigma (= \frac{\pi}{\lambda} E)$. The amplitude of the incident wave becomes (Cowley 1992)

$$\phi(x, y) = \exp(-i\sigma V(x, y)) \tag{7.5}$$

You might have recognised this as the familiar exponential form which expresses phase changes.

Equation 7.5 represents the transmission function of the specimen material. This is called phase object approximation (POA) which tells us what changes in phase the incident wave suffers as it travels through the thickness (see Fig. 7.25) of a material (of the specimen). This is called an approximation because we are neglecting the

Fig. 7.25 Perfect phase object

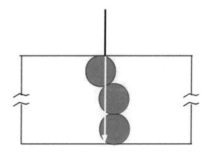

scattering of the incident beam in x,y-directions. Absorption can also be considered in a similar way as a 3-D parameter whose projection in the z-direction would be $\mu(x,y)$ and the transmission function takes the form:

$$\phi(x, y) = \exp(-i\sigma V(x, y) - \mu(x, y)) \tag{7.6}$$

This is only a representation of phase changes as the beam traverses the thin specimen and fails in the case of thick specimens. The exponential term signifies the non-linear addition of phases. With the help of the transmission function, we can also write an expression for the amplitude of an exit plane wave function. This will be of the form $\xi(x,y)$ (note it is 2-D) which equals the product of incident wave amplitude $\xi_o(x,y)$ and the specimen transmission function ϕ (x,y). The objective lens forms a diffraction pattern in its back focal plane, the beams from which interfere once again to form a magnified image of $\xi(x,y)$. Such an image of a true phase object, when formed by an ideal microscope (i.e. aberration-free), would show no contrast under focussed condition. Different ways are available to regain contrast. One of them is to change the focus of the objective lens (you may recall the discussion on depth of focus), either under or overfocus. The other method is to take into account the aberrations associated with the objective lens in formulating the transfer function. In any case, the image is not directly interpretable even if we know the crystal structure. The non-linear exponential term in Eq. 7.6 can be avoided by making another approximation as far as the object is considered. If the specimen is very thin, as is considered to be, the projected potential V(x,y) will be much less than unity. Under these assumptions, the exponential term can be replaced by the following, considering only the first two terms of the expansion of the exponential.

$$\phi(x, y) = 1 - i\sigma V(x, y) \tag{7.7}$$

after neglecting the absorption term. This is known as weak-phase object approximation (WPOA). The second term $i\sigma V(x,y)$ represents weak scattering amplitude and the first term 1 represents the transmitted beam whose amplitude is unaffected during transmission (kinematical condition). Till this stage, we have seen only the transmission of the incident electron wave by the phase object. We now need to formulate an expression for the amplitude of the scattered waves in the back focal plane of the objective, which interfere further to give rise to an image down the optic axis. An ideal microscope behaves like an information processing system where input signals are linearly combined and delivered by the system. A point in the object should be delivered to a point in the image; therefore, a microscope in reality transfers a point in the object to a disc in the image even without considering magnification, as shown in Fig. 7.26. Mathematically, the function which represents this spread is known as the point spread function p(x,y).

The exit plane wave from the specimen $\xi(x,y)$ will then be convoluted with the point spread function by the objective lens to give rise to image wave function $\xi_{image}(r)$ as

$$\xi_{image}(r) = \xi(x, y)...P(x, y)$$

Fig. 7.26 A point object
spreads as a disc

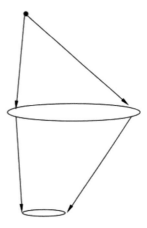

It has been shown in Fig. 4.7 (Sect. 3 of Chap. 4) that exit waves from the bottom
surface of the specimen are brought to focus at its back focal plane by the objective
lens forming a diffraction pattern. When the waves travel further down the optic
axis, they interfere once again to form the image. In the phase contrast image, these
waves are parallel rays (since we assumed very small scattering angles) that meet at
an infinite distance. The objective lens fulfils this requirement because it is making
these rays interfere at the back focal plane. Therefore, further propagation of these
rays from infinity back to the image plane is the inversion of the interference event.
That is the reason for mathematically stating that the inverse Fourier transform of
an image is the diffraction pattern and an inverse of it will give rise to the exit plane
wave amplitudes leaving the specimen bottom surface.

Without further discussion on these lines, we now proceed to the application of
these ideas to explain imaging of a weak-phase object (interested readers are referred
to Spence (2013) and Cowley (1992) for a detailed description). The image wave
function is obtained by convoluting the objective transfer function by the spread
function of the objective, i.e.

$$\xi_{image}(x, y) = (1 - i\sigma V(x, y)) \otimes P(x, y) \tag{7.8}$$

The spread function of an objective lens is a complex number and can be expressed
as

$$P(x, y) = Cos(x, y) + iSin(x, y)$$

The convolution is carried out only with the imaginary part of the transfer function
so that the phase changes associated with the scattered waves are able to modulate
the constant term 1 and generate image contrast. Intensity distribution in the image
can be computed by multiplying the amplitude term by its complex conjugate.

$$I(x, y) = |\xi_{x,y}|^2 = 1 + 2\sigma V(x, y) \otimes Sin(x, y) \tag{7.9}$$

where Sin(x,y) can be obtained by the Fourier transform of the imaginary part of lens transfer function. We can simplify the equations by defining a transfer function of the lens, which is also called coherent Transfer function, as

$$T(U) = A(U)Sin\chi(U) \qquad (7.10)$$

where A(U) refers to aperture function that has a value of unity for all the values of U which are less than the diameter, 'a' of the objective aperture in real space and $\chi(U)$ is the phase factor. The phase distortion function can be adjusted or tuned either by changing the focal length Δf of the objective lens or tuning its aberrations. This function can be derived from fundamental principles (you may consult Spence 2013 or de Graef 2003) as the following:

$$\chi(U) = \pi \Delta f \lambda U^2 + \frac{\pi}{2} Cs\lambda^3 U^4 \qquad (7.11)$$

where U refers to the spatial frequency which is reciprocal of the interplanar spacings of real-space atomic planes and Cs is the spherical aberration constant. In the above simplified expression, we have considered phase distortions due to spherical aberration only as this aberration is present to some degree even in axial rays also, however small it may be (Modern microscopes have better control over accelerating voltage and hence on chromatic aberration). The rate of variation of $Sin(\chi(U))$ as a function of U can be very different from that of normal $Sin(\theta)$ since $\chi(U)$ is a function of two variables in this case. One of them, Δf, can take either +ve or −ve values and the other U itself, which is in quadratic and quartic powers. A typical curve for the transfer function T(U), calculated for a microscope operating at 200 kV with Cs value of 1.2 mm and Δf −90 nm (equal to 50 steps of the fine-focus knob, 1.8 nm each), is shown in Fig. 7.27 along with the spatial frequencies of Aluminium superimposed on it.

For U = 0, T(U) starts at 0 and slopes down till −2.The curve continues and crosses the x-axis at 2.42 $(nm)^{-1}$. It means that for all spatial frequencies beginning from infinity to 2.42 $(nm)^{-1}$, the entire spectrum is transferred with varying dark contrast. The first zero at 2.42 $(nm)^{-1}$ also reflects a very poor resolution limit of the microscope ($\Delta r = 0.413$ nm). We have superposed spatial frequencies corresponding to the first two reflections of aluminium, which are lost in the wildly fluctuating $Sin(\chi(U))$ curve. This behaviour is due to an incomplete balancing of the spherical aberration effect by the −ve sign of Δf (i.e. under focussing of the objective lens). Under focussing takes the beams to focus beyond the specimen. Ideally, we desire a transfer function of the type shown in Fig. (7.28) which has a square trough up to the closest spacings in the material (largest U values). The second zero in T(U) is where the curve crosses the x-axis the first time. At this spatial frequency, the contrast again would be zero, meaning thereby that this spatial frequency is not transmitted by the Transfer function. We notice in Fig. 7.27 that in the second half of the first cycle the amplitude did not reach its maximum (at 'A'). If we push this down to the negative side by tweaking the value of Δf, then the width of the trough can be increased, and

Fig. 7.27 T(U)function calculated for a microscope operating at 200 kV, with Cs = 1.2 mm and at $\Delta f = -90$ nm

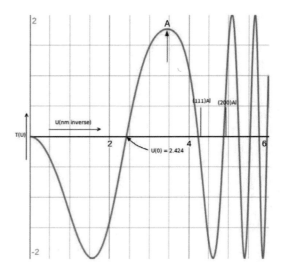

Fig. 7.28 An ideal transfer function

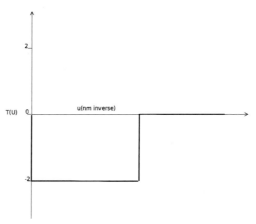

the first crossing over may take place close to 5 $(nm)^{-1}$. An analytic approach can be followed instead of making arbitrary choices, by taking the derivative of T(U) with respect to U and equating the same to zero, so that we know where the Sine function can be nearly flat.

Thus,

$$\frac{\partial \chi(U)}{\partial U} = 0 = 2\pi \Delta f \lambda U + 2\pi Cs\lambda^3 U^3 \tag{7.12}$$

as also

$$\Delta f(0) = -Cs\lambda^2 U_1^2 \tag{7.13}$$

where U_1 corresponds to the centre of the passband. In addition, we desire that $Sin(\chi(U))$ is as close to -1 as possible so that uniform dark contrast is seen for all

Us.

$$i.e. \; \chi(\Delta f_o) = \frac{-\pi}{2}(1, 5, 9, 13...)$$

for negative maxima (i.e. positive contrast).

In general, this occurs for all values of n where

$$\chi(\Delta f_o) = \frac{-\pi}{2}\left(\frac{8n+3}{2}\right) = \frac{-\pi}{2}(Cs\lambda^3 U^4) \tag{7.14}$$

since the second term of $\chi(U)$ equation containing spherical aberration term is balanced by the defocuss term. Therefore, from Eq. 7.14,

$$U^2 = (Cs)^{\frac{1}{2}}\lambda^{\frac{-3}{2}}\left(\frac{8n+3}{2}\right)^{\frac{1}{2}} \tag{7.15}$$

By the substitution of Eq. 7.15 in Eq. 7.13, we have

$$(n) = -Cs\lambda^2\left[Cs^{\frac{1}{2}}\lambda^{\frac{-3}{2}}\left(\frac{8n+3}{2}\right)^{\frac{1}{2}}\right]$$

Extended regions in spatial frequencies over which the $Sin\chi(U)$ function and consequently the Transfer function $T(U)$ is flat (Passband) and the contrast remains dark over these spatial frequencies can be found for many focus settings, i.e. Δfs as

$$(n) = -Cs\lambda\left(\frac{8n+3}{2}\right)^{\frac{1}{2}} \tag{7.16}$$

For n = 0, $\Delta f(0) = -Cs\lambda\left(\frac{3}{2}\right)^{\frac{1}{2}}$

$$i.e. \; \Delta f(\text{Scherzer defocus}) = -1.22(Cs\lambda)^{\frac{1}{2}} \tag{7.17}$$

is known as the Scherzer defocus as defined by Scherzer way back in 1949. The U value corresponding to this Scherzer defocus, where $Sin\chi(U)$ crosses the first zero, defines the point resolution of the microscope since the curve rapidly oscillates after that.

$$0 = \pi(\Delta f(0)\lambda U^2 + \frac{\pi}{2}(Cs\lambda^3 U^4)$$

$$i.e. \; 0 = -\sqrt{(\frac{3}{2})}(Cs)^{\frac{1}{2}}\lambda^{\frac{1}{2}} + Cs\frac{\lambda^2 U^2}{2}$$

upon substitution of $\Delta f(0)$.

Fig. 7.29 A real passband

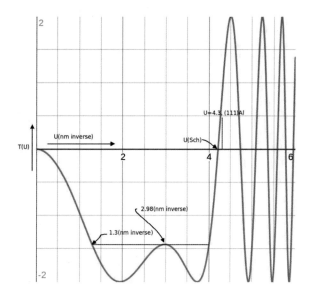

Therefore,

$$U(Sch) = 1.56(Cs)^{\frac{-1}{4}} \lambda^{\frac{-3}{4}} \tag{7.18}$$

and

$$\Delta r(Sch)(Point resolution) = 0.64Cs^{\frac{1}{4}} \lambda^{\frac{3}{4}} \tag{7.19}$$

Equation 7.19 can be derived in a different way as well.

Real Passband: Using the expression for δf (Eq. 7.13), we can compute the Scherzer defocus value knowing the Cs value of the microscope being used (usually provided by manufacturers as a specs.) and operational voltage, which is usually 200 kV. The T(U) plot shown in Fig. 7.29 is for a microscope operating at 200 kV, with Cs of 1.2 mm and $\Delta f(0) = -67$ nm. The first passband occurs near the low spatial frequency end of the spectrum. As desired, the central U value of the passband is at 2.97 $(nm)^{-1}$ with a T(U) value of -1.435, though not the maximum achievable value of -2.

Therefore, the microscope transfers all the frequencies from 1.3 to 4.0 $(nm)^{-1}$ with the same dark contrast, rendering an interpretable image. The U value where the curve first crosses the x-axis defines, as earlier, the U value for which the contrast is zero (or the least, as other factors give rise to some contrast) and least detectable. A through-focal series of images of amorphous holy carbon can be taken where the least contrast position can easily be detected. Such a series is shown in Fig. 7.30. Only two images on either side of $\Delta f = 0(c)$ are shown in (a) and (b) for want of space.

This U value defines the point resolution of the microscope since the image is most easily detectable during a change of Δf by the microscopist. There are, of course, other definitions of the limit of resolution. The U(Scherzer) works out to be

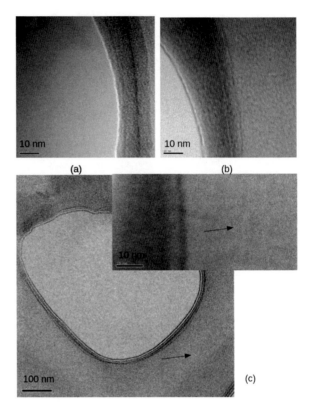

4.21 $(nm)^{-1}$ corresponding to 0.24 nm real space. The Aluminium (111) spacing
falls just outside the passband. Although the first passband terminates at U(Sch), we
can compute the information from the subsequent −ve amplitudes up to some more
cycles beyond which the cycling is very rapid. The U value up to which we can do
this is known as Information Retrieval Limit.

The general expression for passbands gives us a way out from this. By tuning
T(U) function, we can get a passband at the spatial frequencies of our interest as
shown in Fig. 7.31. Here, the passband is tuned by setting the Δf at −230 nm and
Cs = 1.2 mm for transferring (200) Aluminium spacing. Remember that such tuning
cannot go *ad infinitum*. Think of the reasons, why it is so!

Actual transfer function in a microscope can be very different from the ideal transfer
function plots that we discussed above. Often, the higher order aberration terms form
an envelope over the transfer function and attenuate it completely at larger spatial
frequencies beyond the first passband. The factors which are responsible for the
envelope function are the lack of spatial coherence of the incident beam and the
chromatic aberration effects. The Transfer function can be expressed as below in
such cases.

$$T_{actual}(U) = T(U)EaEc \qquad (7.20)$$

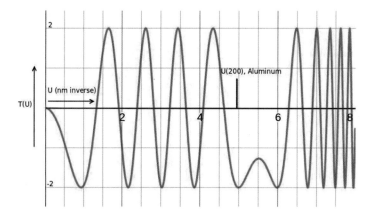

Fig. 7.31 A passband tuned to occur at the desired spatial frequency

The envelope functions affect the A(U) function by constricting it.

The envelope function due to the spatial coherence of the beam is gun-dependent and is much better in field-emission guns. The information retrieval limit is further beyond the point resolution defined by the Scherzer defocus. A schematic $T_{eff}(U)$ is shown in Fig. 7.32a.

After aligning the microscope and choosing the right parameters for imaging, we proceed to record either the lattice fringes or structure images depending on the limit of resolution of the microscope and the spatial frequencies available in the material that we are investigating. Figure 7.32b shows lattice fringes of indium islands at the interface between Al and In in a rapidly solidified, phase-separated Al-In alloy (Courtesy: Images were obtained using the HREM facility at IGCAR, Kalpakkam) (S. Umamaheswara Rao, M.Tech. Dissertation, Dept. Met. Engg., IIT-BHU, 2004). Figure (c) shows lattice fringes (rather multiple graphite layers) of a multiwall carbon nanotubes (MWCNTs) (Images taken by the author during the tenure of Dr. R. H. Kulkarni Memorial Visiting Fellowship at RPI, Troy, USA, Host: Professor P. M. Ajayan). The structure image given in (d) is once again that of an indium particle displaying the distinct difference in atomic arrangements (disorder) at the periphery (i.e. growth front) and interior (ordered and crystalline). Yet another image of lattice fringes is given in Fig. 7.33a. These are fringes of iron oxide nanoparticles which are very thin. Salt-pepper contrast of the underlying support carbon film can be noticed clearly. Further, the fringes show dotted contrast, a structure image that is barely resolved. The true amorphous structure of the support film can be noticed (Say at A) indicating a perfect correction of the objective astigmatism. The first fringe shows a reversal of contrast that needs a deeper understanding of contrast by modelling. Figure 7.33b shows an overlap of fringes due to two such particles superposed on each other. At the interface, once again an apparent structure image is seen (at white arrows) which is not a true resolution of atomic spacings of the structure in 2-D. We need to exercise caution in interpreting such images.

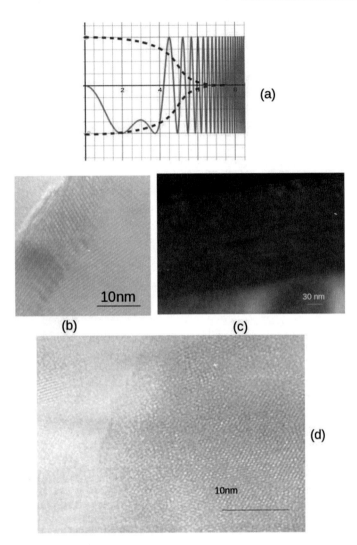

Fig. 7.32 **a** A realistic transfer function. **b** Lattice fringes at the interface between Al and In phase-seperated regions. **c** Fringes in MWCNTs. **d** Structure image of an indium particle. (Micrographs (**b**) and (**d**) are provided through the courtesy of IGCAR, Kalpakkam. Images were obtained using the HREM facility at IGCAR, Kalpakkam, while the author was on a BRNS Visiting Fellowship. With permission) (after S. Umamaheswara Rao, M.Tech. Dissertation, Dept. Met. Engg., IIT-BHU, Varanasi, India, 2004. With permission)

Fig. 7.33 a Lattice fringes of iron oxide nanoparticles. **b** Overlapping Lattice fringes at the interface between two such particles giving an impression of a structure image. (after Pankaj Goel, M.Tech. Dissertation, Dept. Met. Engg., IIT (BHU), Varanasi, India, 2004. With permission)

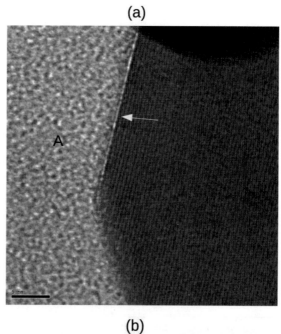

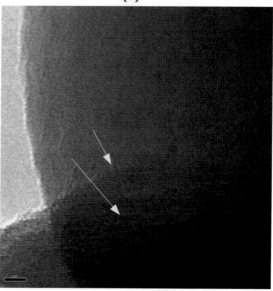

7.4 Differential Energy Contrast

We assumed in a kinematical theory that the incident loss-electrons are either mostly transmitted or weakly diffracted by a thin specimen without undergoing any energy loss. Nevertheless, some of the incident electrons interact with the plasma and the inner shells of the atoms and suffer losses. Given the very small energy spread of the present-day electron guns, field-emission guns in particular, it is possible to resolve the energy losses suffered by the beam electrons as they pass through the specimen. These inelastic interactions form the basis of electron energy-loss spectroscopy (EELS) (Fig. 7.34) which acted as a precursor to energy-filtered imaging. The spectrum of energy losses could be converted to form an image or a diffraction pattern by suitably modifying or adding additional components to the spectrometer and obtain energy-filtered images. Most of the modern analytical electron microscopes come with this configuration. The Energy-Filtered Transmission Electron Microscope (EFTEM) gives images that show contrast due to differential energy. The range of energy values that the loss-electrons represent is from 0 to 3000 eV. EFTEM imaging is referred to as energy-filtered imaging (EFI) or electron spectroscopic imaging (ESI) by some authors. The energy values that we are referring to are indeed differential energies or energy differences. In the nomenclature of EELS spectroscopy, these are referred to as energies corresponding to absorption edges and so on. EFTEM can be performed in a transmission electron microscope that is suitably equipped. We need to keep the energy slit small enough to minimise chromatic aberration (Please recall from our discussion about chromatic aberration from Sect. 6.3.2 that chromatic aberration in TEM is due to the energy spread in the accelerating potential of the electron gun that manifests in the objective and the image. Note that the objective lens collects all the beams from the specimen beforehand. In the context of EFTEM, we are dealing with an image that forms due to the energy spread and hence the concern about chromatic aberration) and select an electron beam with a specific energy. This selection can be made by slightly lowering the gun

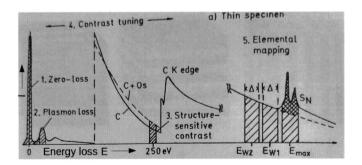

Fig. 7.34 A typical EELS spectrum. (Figure 4.30a provided through the courtesy of L. Reimer and H. Kohl 'Transmission Electron Microscopy: Physics of Image Formation', Fifth Edition, Springer Science+Business Media, LLC, 2008. With permission)

energy and bringing the desired beam close to the axis. We note that there are two types of energy filters available from the manufacturers; the in-column (Ω) filter and the post-column filter (GIF).

In-column filters are placed between the intermediate lens and the projector lenses such that the recording CCD detector only receives electrons that came through the filter. So all images/DPs consist of electrons of a specifically selected energy. We can, of course, turn off the filter and continue with the normal operation of the TEM. Though it is possible to record an EELS spectrum, it is not recommended in this configuration.

The Ω filter: These are a better substitute for the mirror prisms used in the erstwhile models. The filter consists of a set of magnetic prisms arranged in the form of an Ω, which disperses electrons to different angles on the basis of their energy, but in the end of the path brings them back on the axis of the microscope before entering the final projector lens. Note that the prisms have peculiarly shaped end-faces (as illustrated in Fig. 7.35) which are machined so, to reduce aberrations. The procedure followed

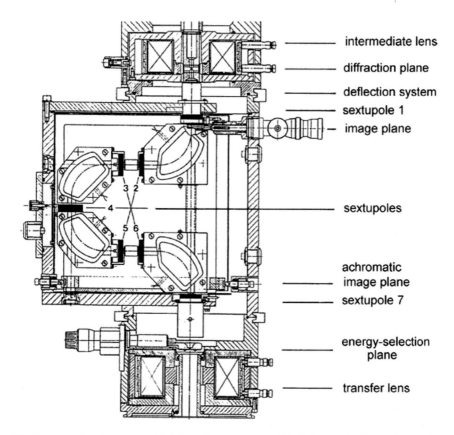

Fig. 7.35 Ω filter (Figure provided through the courtesy of L. Reimer and H. Kohl 'Transmission Electron Microscopy: Physics of Image Formation', Fifth Edition, Springer Science+Business Media, LLC, 2008. With permission). Original figure to be obtained by Springer for reproduction)

to acquire images in EFTEM is as follows: We usually project the diffraction pattern into the back focal plane of the intermediate lens. Therefore, the entrance aperture of the spectrometer selects an area of the specimen, and β the collection angle is then governed by the objective aperture. Electrons following a particular path (energy selection) are selected by the post-spectrometer aperture. Thus, electrons of a chosen energy range determined by the width of the aperture (variable) and strength of the magnetic prisms, usually those that suffered either only plasmon-loss or particular edges (e.g. k, etc. in the EELS spectrum, Fig. 7.34) or a mixture are used to form the image onto the CCD camera of the TEM. Several sextupole lenses, are added to this specially designed Ω filter in F-H Institut, Berlin, to take care of the aberrations in the optics of the Ω filter.

In a similar manner, we can tune the TEM optics to project a diffraction pattern into the prism entrance aperture which then produces an energy-filtered DP (or even a CBED) on the CCD Camera. If we then select a portion of the DP, we can get an energy-loss spectrum displayed on the CCD.

Post-column Gatan Image Filter: The post-column energy filter, usually installed on all machines, is from GATAN which consists of an energy-selecting slit, a magnetic prism and a 2-D slow-scan CCD camera. The magnetic prism used is a sectorial magnet followed by many multipole lenses. The post-acquisition components are more complicated in comparison to an in-column filter (not the F-H-I one) as shown in Fig. 7.36. The central electron beam in a symmetric magnetic prism follows a

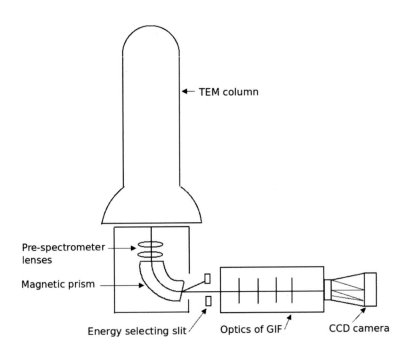

Fig. 7.36 Post-column filters (GIF)

Fig. 7.37 Focussing action

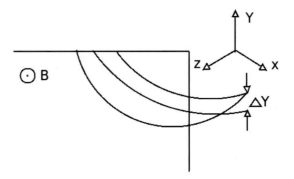

circular path through the prism with radius $r = \frac{mv}{eB}$ where B is the magnetic field, m is the mass of the electron, v is the velocity and e the charge. It passes out from the exit face that lies at right angle to the entrance plane. Electrons with other momenta are focussed to a point ΔY above this point (see Fig. 7.37). Therefore, a slit placed at this location can pick up the desired spectral line. Note that there is no focussing in the z-direction parallel to the magnetic field which leads to second-order aberrations. They can be corrected by curving the entrance and exit faces of the magnetic prism. If we go into finer details, we should also expect the magnetic field to extend beyond the faces and need to be taken care of. The energy resolution of the spectrometer largely depends on the width of the image of the entrance slit and second-order aberrations. However, even with thermionic guns, resolutions of the order of $1-2$ eV can be achieved. Further corrections of the aberrations are carried out by the sextupole and octupole lenses within the GIF optics so as to achieve resolutions of the order of 1 eV. Such resolutions also require a FEG source for the electron gun. Fine-tuning of sextupoles and octupoles is not a manual job, but needs a controlling software that is embedded in the system software. Magnification of GIF also needs to be adjusted such that (Note that GIF is below the viewing screen of TEM) the actual TEM magnification is reduced considerably in order to see the image on GIF with a reasonable magnification. Recent models of the instrument overcome this drawback.

Energy-filtered imaging or diffraction has many advantages in terms of the clarity of contrast, knowing precise elemental distribution in precipitate phases such as inter-metallic carbides and nitrides, zero-loss filtered diffraction patterns from crystalline or amorphous phases and even for biological specimens for obtaining structure-sensitive contrast. Both post- and in-column energy filtering were very popular in the 80s and 90s. In a way, the high angle annular dark field imaging (see Sect. 7.5) in STEM has taken over now. Interested readers are referred to Sect. 4.6 of Reimer and Kohl (2008) and Chap. 37 of Williams and Carter (2009).

7.5   Contrast Mechanisms in STEM

A scanning transmission electron microscope (STEM) uses a small electron probe to scan the sample and form an image on the detector, pixel by pixel using the signals transmitted by the specimen. A choice can be made from the many signals emitted by the specimen and operate STEM in that mode. At present, STEMs are available as an attachment to a conventional electron microscope in addition to stand-alone models. Its availability as an attachment to TEM is a consequence of the fact that components and ray paths in the two instruments are common and are related to each other by what is known as *reciprocity theorem* which is applied in optics, electrical networks, etc. A schematic ray diagram of STEM is given in Fig. 7.38a. Also presented alongside in Fig. 7.38b is the ray diagram of a typical TEM for understanding the applicability of the reciprocity theorem. In the context of a TEM, the theorem can be understood that under elastic scattering conditions if the source of electrons and detector are interchanged in their position in TEM, the ray paths remain unaltered. This can easily be realised by observing the ray paths within the boxed region in the figure. The same principle is stated in a lighter way some times that the STEM optics are that of an inverted TEM giving scope to imagine that starting from the electron gun to the detector are in reverse order of their location in TEM, but we notice that the STEM detector is located below the observation screen of TEM. Importantly what the applicability of the reciprocity theorem means for us is that the diffraction contrast (i.e. image contrast) and phase contrast in a STEM image can be interpreted more or less in the same manner as in TEM.

Contrast: We will not discuss the usual diffraction contrast in STEM images here but take up only the phase contrast images. We have seen from Fig. 7.39a that the specimen is scanned by a fine convergent probe. Therefore, we can expect a convergent beam diffraction pattern with overlapping discs on the entrance slit of the STEM. Interference of the waves takes place at the overlap regions of the diffraction discs as shown in Fig. 7.39. Further addition of beams takes place as the probe scans several points of the specimen. They interfere to give rise to phase contrast in the image. In order to interpret the image, we can take the help of the Transfer function that was discussed in Sect. 7.3 and represented in Eq. 7.10. We need to show that it is valid in the case of STEM as well, provided that the objective aperture is filled with coherent electrons, as assumed in that section. A convolution of the probe on the specimen then becomes the inverse Fourier transform of the electron wave at the front focal plane (towards the gun). The specimen diffracts the incident probe giving rise to discs (diffraction) mentioned in the schematic diagram. We take a Fourier transform once again to formulate the expression for the electron wave reaching the detector plane. The Fourier transform would show that at the detector space the observed 'image' is a diffraction pattern in reciprocal space with overlapping discs in case the objective aperture is large.

Upon computing the intensity at the overlap regions, we notice that it is related to the interference fringes. The sharpness or resolution of these fringes may also get

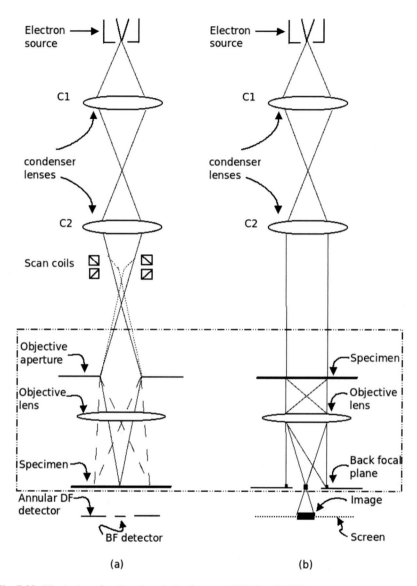

Fig. 7.38 Illustration of reciprocity relation between STEM and TEM

affected by the objective lens aberrations. The fringes can be visualised as shadow images of the lattice planes in the diffraction space (recall the discussion on LACBED in Sect. 6.4).

BF image: High-resolution images can be obtained in STEM both in BF and DF modes. We need to align the detector on the optic axis for BF imaging. Intensities that contribute to the image are the zero, $+\mathbf{g}$ and $-\mathbf{g}$ beams. We will observe that the

Fig. 7.39 STEM imaging

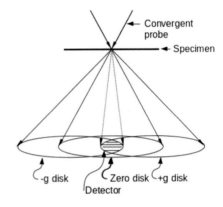

Fig. 7.40 BF-DF STEM detector

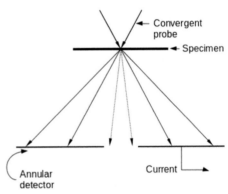

expression for the intensity is similar to that formulated for a phase contrast image in TEM. It should be noted that the intensity in the image is much less than in the HREM images of TEM since the BF detector collects only a small portion of the electrons reaching the detector plane (see Fig. 7.40). This drop in intensity is also due to the diminished probe size as a result of the combined effect of the focussing condenser and objective lenses. Resolution is dictated by the condition that $|g|$ should be less than the radius of the on-axis aperture.

Dark field or annular dark field imaging: An annular detector is designed to collect the diffracted beams through the central opening, leaving out the direct beam. The surface of this thin annular detector is electron sensitive and collects the signals coming at an angular range of a few tens of milliradians to 100mrads, as shown in Fig. 7.40. Some researchers opine that at large scattering angles the signal is incoherent, largely due to thermal diffuse scattering (TDS) by phonons or the Rutherford scattering from core electrons and hence should be dealt with separately as high angle annular dark field (HAADF). We will first consider the coherent elastic scattering, i.e. at the Bragg angles to form the ADFs. The intensity of the image can be calculated by considering the Fourier transform of the probe intensity and convolving it with the Fourier transform of the object function, whether continuous in **k**-space

Fig. 7.41 Optical transfer
function

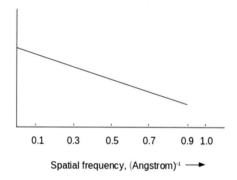

or discrete in **k**-space (i.e. crystalline specimen). In the case of STEM, the Fourier
transform of probe intensity is a monotonically decaying function with respect to
spatial frequency (see Fig. 7.41). The greatest advantage of dark field STEM imaging at high resolution lies in this fact. The Fourier transform is known as the optical
transfer function and is equivalent to the transfer function of an objective lens in
TEM (recall Eq. 7.10 from Sect. 7.3). The OTF shows no oscillations and passbands.

Therefore, the observed contrast is directly interpretable. Since a square term
is involved, the −ve T(U) values in the case of TEM also become positive, and
hence, the contrast transferred is the same at all spatial frequencies. Moreover, there
is no delocalisation effect in this incoherent image. The bright dots seen in the
high-resolution image can be directly correlated to atoms or atomic columns. If an
online spectrum is also obtained from the same location of the specimen, then the
compositional assessment can also be done. Since the contrast is directly related to
the atomic number of the atoms, the larger the 'Z' value the higher is the contrast.

HAADF image contrast: The structure image of a crystalline specimen is a projection of the inter-atomic potential, as we discussed in the case of TEM. This condition
is still valid in the case of STEM imaging although we form a highly focussed beam
on the specimen, unlike the parallel beam used in TEM for structure imaging. The
atomic columns, nevertheless, 'see' a beam that is two orders of magnitude larger in
diameter. At any instant of time, the scan interacts with only one column of atoms
when the probe is stationed over it and electron channelling takes place parallel to
the column. When the probe size is smaller than inter-column spacings of atoms, the
probe proceeds picking up signals from columns sequentially during the scan. Thus,
there is no coherent interference from neighbouring columns. The signals reaching
the detector remain incoherent and elastically scattered. Owing to this, image contrast remains independent of the structure factor effect, etc. Of course, in saying
so, we are assuming that there is no beam spread—an assumption valid only in the
case of thin samples. There is considerable influence of thermal diffuse scattering
(TDS) in thick samples and at larger scattering angles. TDS is a phenomenon arising
from the interaction of the electrons scattered by a column with the quanta of lattice vibrations. They scatter electrons incoherently but elastically. Further, the probe
while dwelling on an atomic column also interacts with the core electrons giving rise

Fig. 7.42 a Shows a normal
BF STEM image of the
Co-Fe-Mn oxide along with
the corresponding diffraction
in the inset. **b** HAADF image
of Co0.6Fe0.8Mn1.6O$_4$
spinel system after sintering
at 1250 °C for 24 h followed
by ageing at 350 °C for 250
h. The Dark and bright
contrast of striped
nanodomains is signifying
the Mass-thickness contrast
arising due to phase
separation of Fe-rich and
Mn-rich phases. The Fe-rich
and Mn-rich phases turn out
to be CoFe$_2$O$_4$ and
CoMn$_2$O$_4$, respectively
(Micrographs provided
through the courtesy of
Avnish Singh Pal and
Joysurya Basu, 2017,
Unpublished work,
Department of metallurgical
Engineering, IIT (BHU),
Varanasi, India. With
permission)

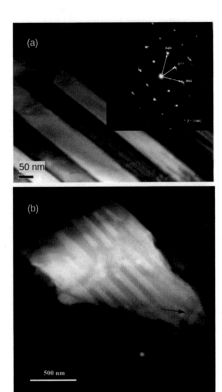

to the Rutherford scattering. The effect of phonon scattering on the image contrast
needs to be simulated for any quantitative interpretation. The inelastically scattered
electrons from single columns can give rise to very high-resolution images and are
made further sharp by the aberration-corrected STEMs. A HAADF image of magnetic domains in Co-Fe-Mn oxide is given in Fig. 7.42b for illustration. Figure (a)
of it shows the normal BF STEM image with an inset showing the corresponding
DP. Note the presence of two sets of domains in the HAADF image in the lower part
(indicated by the arrow).

7.6 Magnetic-Moment Contrast

While imaging magnetic materials, which are either ferromagnetic steels or magnetic
thin films, we experience many difficulties in aligning the beam along the optic axis
of the microscope or in obtaining sharply focussed images as a result of deflecting
force exerted by the magnetic field of the specimen on the electron beam. So much
so that we might find it tedious to work with magnetic specimens! We must define
our goal in such a case, whether it is to image magnetic domains in the specimen or

to image the microstructure of a magnetic material (such as a specimen of magnetic steel). The two procedures are different.

The special steps to be taken for examining magnetic specimens begin from the sample preparation stage itself. Small pieces of specimens cut from bigger coupons that are thinned by *window* technique are preferred to jet polished or ion-milled 3 mmϕ discs as the latter can exert a magnetic field of $\frac{1}{5}$T. If the samples are in powder form, we need to prepare specimens by embedding them in a contrastless polymer such as Formvar or Collodion. Otherwise, the powders jump off the grid and stick to the pole pieces. In all the above cases, we first reduce the objective lens current to a considerable extent before introducing the specimen holder to the microscope. We should not adjust the specimen height to eucentric position at this stage. Some users practise switching off the objective lens and using the 'low mag' mode available in the microscope. However, do not switch off the objective without first checking whether the other lenses are aligned with respect to C2. This method may not also suit the requirements of magnetic materials with ultra-fine-grained microstructure. After locating an area of interest, we begin tilting the goniometer to only one side of its zero and try to reach the eucentric position in small cycles of tilt and 'un-tilt' without ever crossing the zero tilt position of the goniometer. We remove the objective aperture and reduce the current in the objective and condenser lenses significantly by watching the display of currents on the system monitor. The image should be centred using only the DF tilt controls.

Now we can refocus the image and energise the C2 lens such that the image is as sharp as it could be. We can check whether the transmitted beam is in the centre of the screen in DP mode. After repeating the above-mentioned steps a few times and confirming that the transmitted beam is at centre of the screen, we place the objective aperture around the central beam and switch to image mode. Figure 7.43 shows a BF image of a bainitic steel obtained in this manner.

Note that all the above settings are valid only for the region for which they were carried out. If the specimen is translated, the image will again go off the screen (actually the beam) and the above-mentioned steps need to be repeated.

7.6.1 Lorentz Microscopy

When the study of the magnetic domain structure of the specimen is our goal, we adopt a technique called the Lorentz microscopy, since the domain structure is not resolved in the normal microscopy. The principle associated with this technique is to calculate the force exerted by the magnetic specimen on the incident beam, which is called the Lorentz force. The beam gets deflected by an angle θ since this force acts in the plane of the sample. By a calculation of the Lorentz force (deGraef 2003), we can arrive at the deflection angle θ as

$$\theta = \frac{e\lambda}{h} B \perp t \qquad (7.21)$$

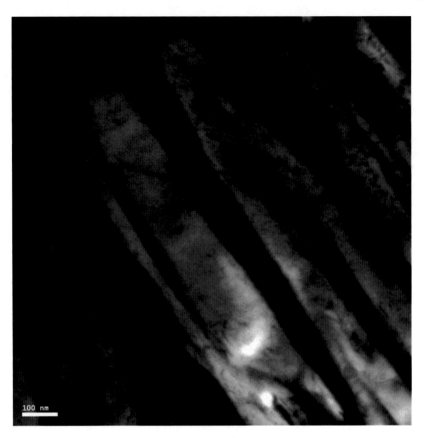

Fig. 7.43 BF image of a bainitic steel. Note the lack of sharpness in the image (after P. Madhavi, M.Tech. Dissertation 2019, Department of Metallurgical Engineering, IIT (BHU), Varanasi, India. With permission)

where e is the charge of the electron, λ the wavelength corresponding to the accelerating voltage used, h Plank's constant, $B\perp$ is the component of magnetic induction perpendicular to the beam and t the thickness of the specimen traversed by the beam (i.e. local thickness). Since θ is dependent on t, it was recommended earlier to take a thin specimen for investigation. The lower limit for t is such that it does not alter the domain structure of the specimen due to the influence of magnetic induction from pole pieces of the objective lens.

In the configuration of microscopes that enable the Lorentz microscopy, we have either a separate Lorentz lens or a mini-lens on the objective. Essentially, there are two modes in the Lorentz microscopy—the Fresnel mode and the Foucault mode. We will discuss the Fresnel mode in some detail here and leave the Foucault mode for interested readers to refer to Graef (2003) and Williams and Carter (2009). Consider a hypothetical specimen with three 180° domain walls in it. In the first and the last,

Fig. 7.44 Fresnel mode of
imaging magnetic domains
(on the pattern similar to the
one given in de Graef 2003)

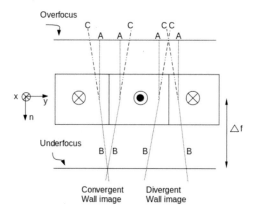

the magnetisation direction points into the plane of paper, while the middle one has it out of the paper as shown in Fig. 7.44. Thus, the three domain walls are 180°-rotated w.r.t. magnetisation.

Imaging with a normal objective lens would result in beams that pass parallel to the domain walls (AA) without 'seeing' them and hence won't render them in the BF image. Under focussing the objective lens strongly increases the focal length and results in a convergent beam in the left side pair (marked BB). Consequently, at the projected location of the domain wall, more electrons get focussed leading to a bright line in the image. Whereas in the right side, domain wall rays marked BB go divergent resulting in a deficit of electrons in the focal plane. This forms a dark line at the projected location of the domain wall. Since the images are obtained by under focussing the objective lens, the images are called the out-of-focus or Fresnel fringes! You can observe from the schematic diagram that the sign of bending of the rays for outer domains is towards the +ve direction of y, which is a perpendicular component of the Lorentz force. Exactly the reverse contrast would be obtained for the overfocus condition.

7.7   Electron Holography Contrast

In high-resolution imaging, we discussed the mechanism by which phase shifts are introduced in a parallel beam of electrons when they pass through a thin specimen. Electrons leaving the bottom surface of a specimen interfere with each other owing to their phase differences and give rise to contrast in the image, simultaneously losing their phase information. Holography is a method of reconstructing this phase information from a hologram to get very important information at atomic resolution in different materials and also electric and magnetic domain structures at very high resolution. D. Gabor invented electron holography in 1948 when he was looking for a method of improving the limit of resolution in TEM. There are three varieties of holograms: (i) in-line holography as originally proposed by Gabor, (ii) off-axis or out-of-line holography as developed by Leith and Upatneiks (1962) and (iii) a 3-D transmission holography using lasers, also developed by Leith and Upatneiks. We will explore the first two techniques as these are directly relevant to materials characterisation.

7.7.1 In-Line Holography

Consider a simplified schematic column of an electron microscope where an object point 'O' scatters the incident plane wave of electrons as shown in Fig. 7.45. The scattered and the incident plane waves move forward and interfere to give rise to an interference pattern (which is a diffraction pattern).

This pattern forms at a distance Δz from the object, and if the photographic film or a digital detector is placed at the interference plane, a hologram gets recorded. It consists of concentric rings with decreasing spacing towards edges and a central disc. In the reconstruction part of the method, the recorded hologram is illuminated by a parallel plane wave of the light beam as shown in Fig. 7.46, from the same direction as the electron beam was incident while recording the hologram. Light gets diffracted from the hologram at an angle of $\pm\theta$. The two diffracted waves, also called side-bands, give spherical wave from O' and O". The reconstructed object is O' and its twin image O". The formation of a twin image is the principal disadvantage of in-line holography since the two images and the transmitted beam are collinear. Though we assumed a point object in the above construction, we can easily infer what happens to object points that are above and below point O, as well as in its front

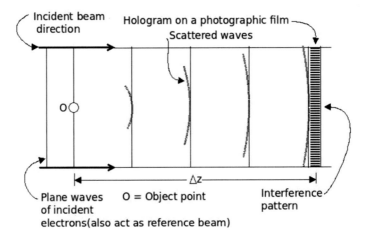

Fig. 7.45 In-line hologram

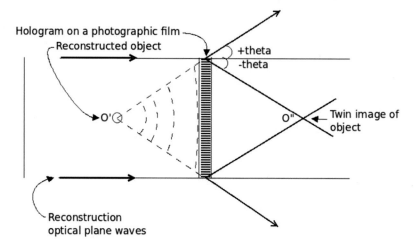

Fig. 7.46   Reconstruction of an in-line hologram

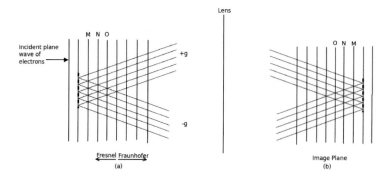

Fig. 7.47   a Fresnel-type hologram. **b** Fraunhofer-type hologram

and rear. The asymmetric holograms will give rise to object points around O' in the reconstructed object to give rise to a 3-D image.

Interference of rays, reference and scattered, may take place either in the near-field (Fresnel) or far-field (Fraunhofer) as given in the schematic diagram, Fig. 7.47a and b their reconstruction, respectively. They can be recorded separately as the Fresnel and Fraunhofer holograms.

No information is available at Zeros in the hologram as at MNOP. Similarly, there are zeros at the Fraunhofer hologram also (see (b)). They occur at $\Delta z = \dfrac{nv^2}{\lambda}$ where $\frac{1}{v} = q$ is the spatial frequency of the object, λ the wavelength of light used for reconstruction and n is an integer taking values 0, 1, 2..... The zeros occur in the transfer function of the objective lens in HREM at the same defocus as Δz, if only defocus is considered in $T(q) = -2\text{Sin}(\chi(q))$. The hologram so recorded at either near-field or far-field is formed by electrons while the reconstruction is done by an optical beam (laser)—a point to be taken into account while designing the reconstruction optics. A lens is placed with a suitable spherical aberration constant for this reason. The reconstruction shows a transfer function that is square of T(q). Recall T(q) for HREM from Fig. 7.30. We noticed that in addition to zeros in the spatial frequency spectrum, there are periodic reversals of contrast whenever T(q) changes sign due to its dependence on a sine function. In the case of a hologram, reconstruction due to the squaring up of T(q) has the benefit of representing all the spatial frequencies with the same contrast, the zeros remain though. In the Fraunhofer hologram, we would like to avoid the superposition of side-bands by choosing a defocus that is at least $\dfrac{d_o^2}{\lambda}$, where d_o is the area of the specimen from where the hologram was recorded and λ is the wavelength of light used. The difficulty in adopting this method is the very large defocus value that is required. A solution to this, which has been used by some researchers, is to block one of the side-bands with a half-plane diaphragm once and repeat the reconstruction with the other side-band by keeping the half-plane diaphragm in the path of the first side-band. Superposition of the two separately recorded holograms gives only amplitude information suppressing the phase information, and subtraction of one from the other gives the missing phase information.

7.7.2 Off-Axis Holography

The limitations of online holography w.r.t. the collinearity of the twin images and the direct beam were proposed to be overcome in the ingenious design of a biprism by Leith and Upatneiks (1962). The biprism device sends a reference beam that superimposes on an off-line wave that carries the specimen structure. The specimen wave is thus modified in amplitude and phase. The biprism, which operates at a voltage of about 300 V and which consists of a filament having a diameter of $\sim$350 nm, is mounted just above the selected area diaphragm of the intermediate lens of the microscope. The lens itself is focussed on to the hologram plane. The schematic drawing is given in Fig. 7.48.

Typically a few hundred fringes of 0.02 nm spacing and an overlap length of 30 nm can be recorded on the hologram. The reconstruction optics again have biprism and uses a coherent light (laser) to illuminate the hologram. The first-order reflection carries the amplitude information which can be collected by an aperture and thus suppress the twin image. Tonomura (Chen et al. 1993) has developed this method to image magnetic flux lines in a superconducting material. The current trend is to reconstruct the object online by reading the hologram formed by the biprism and give the feed-back signal to the microscope on a separate screen. This facilitates the observation of the dynamic behaviour of magnetic domains and flux lines in real time (Chen 1993). Interested readers are referred to Tonomura (1999).

Fig. 7.48 Off-axis holography

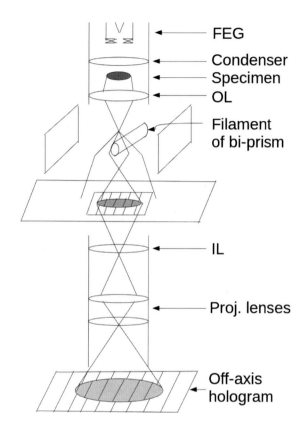

FEG

Condenser
Specimen
OL

Filament
of bi-prism

IL

Proj. lenses

Off-axis
hologram

Exercises

Q1. On which lens does the resolving power of a transmission electron microscope depend? If spherical aberration is having an opposite effect on the limit of resolution (Given the radius of the disc of the image due to spherical aberration $= C_s(\alpha)^3$, where C_s is the spherical aberration constant and α is the angle subtended by the peripheral ray with the optic axis), derive an expression for optimum resolution.

Q2. Find the maximum thickness of a wedge-shaped specimen if the bright field image of it shows three fringes in [111] orientation, given that the extinction distance $\xi_{[111]}$ for aluminium is 55.6 nm. Can there be a possibility that the specimen is pure iron instead of aluminium? Give reasons.

Q3. From the expression for intensity of a diffracted beam under kinematical condition, it is seen that the intensity varies as $I_s = \frac{C^2(Sin^2\pi t s_z)}{(\pi s_z)^2}$. Show that this expression is not valid under the exact Bragg condition.

Q4. What is the contrast expected to be observed in a dark field image when a hypothetical sample has the same thickness as that of the extinction distance for a particular (hkl) plane?

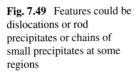

Fig. 7.49 Features could be dislocations or rod precipitates or chains of small precipitates at some regions

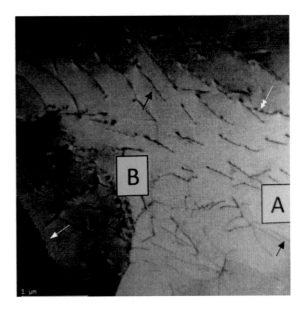

Q5. In a specimen, two phases are present epitaxially such that some crystallographic planes of one phase (that marginally differ in their spacing from those of the other) are parallel to those of the other. What is the contrast observed in the bright field image in such a condition?

Q6. The BF image of a sample is presented in the following Fig. 7.49. The image obtained under kinematical conditions could be interpreted either as (i) small rod-shaped precipitates (marked by a white arrow) or (ii) small-length dislocations in regions marked A and B or (iii) as a chain of small precipitates (marked by the white double arrow). On the basis of observed contrast and on the basis of possible diffraction evidence from these features, identify whether these are precipitates or dislocations. Justify your answer with logical arguments and with schematic figures where necessary.

Q7. (a) Using the kinematical theory of diffraction contrast and the diffraction patterns obtained from the regions corresponding to the cases mentioned below, how do you distinguish between

   (i) an array of straight dislocations vis-a-vis an array of needle-shaped precipitates and

   (ii) stacking faults vis-a-vis twins.

(b) Stacking faults and 1-D defects such as dislocations come in contrast and go out of contrast when the specimen is tilted while keeping the area of observation constant. Explain the reasons.

Q8. Can you observe in a high-resolution image the contrast from a vacancy present in a column of atoms in a crystal? Give reasons.

Q9. Calculate and plot the contrast transfer functions (CTFs) of a TEM with a C_s value of 2.0 mm for different Δf values. Identify the best-suited CTF for transferring the contrast for d spacings in the range of 0.1–0.05 nm among those. Assume the microscope to be operating at 200 kV and the available range of Δfs as $\pm$ 200 nm in steps of 20 nm. Explain any assumptions made in your calculation.

Chapter 8
Lensless Electron Microscopy

A set of three microscopes can be considered as an illustrative list of microscopes which form images of a specimen without the aid of any objective lens. In such microscopes, the optical laws elaborated earlier in Chaps. 2 and 3 will not be applicable. These are a Scanning Electron Microscope (SEM), an Atom Force Microscope (AFM) and an Atom Probe Field Ion Microscope (APFIM) or a 3D Atom Probe. AFM uses the force experienced by a cantilever beam because of attraction/repulsion between a tiny stylus attached to it and the atoms on the surface that it is scanning. The deflection of the cantilever beam is plotted as a depth profile giving a 'pseudo image' of the atoms. The 3D Atom Probe uses atomistically sharp needle of the specimen and subject it to high electric field pulses or laser pulses. The atoms get ionised in the process and evapourate off from the surface so as to land on channel plate or imaging plate. A Time-of-Flight Mass Spectrometer (ToF-MS) coupled to this can map the atomic constituents at the highest resolutions we would aspire for, as it measures the time taken by and the charge on the individual ions. Owing to the high curvature of the electric field generated, the atoms get projected onto a larger area on the channel plate and give rise to requisite magnification (highest) to image atoms. It is perfectly in order to re-emphasise that no objective lens(es) is used in these so-called microscopes to form the image unlike in the case of an optical microscope or a transmission electron microscope discussed in earlier chapters. Although there are many such lensless techniques available at present, we will discuss in this chapter only the one which has the widest applicability in materials characterisation, viz., the SEM.

© The Author(s), under exclusive license to Springer Nature Singapore Pte Ltd. 2022   187
G. V. S. Sastry, *Microstructural Characterisation Techniques*, Indian Institute
of Metals Series, https://doi.org/10.1007/978-981-19-3509-1_8

8.1   Scanning Electron Microscope

Scanning electron microscope (SEM) is a surface characterisation tool for bulk specimens. Its versatility increased with the addition of instrumented chemical analysis tools such as energy-dispersive X-ray analysis, electron backscattered diffraction camera, an adaptable stage that can facilitate either electrical or mechanical measurements in-situ on bulk specimens and an environmental mode where biological wet samples can be observed at elevated specimen chamber pressures, to a basic SEM of earlier times.

8.1.1   A Schematic Layout of the Microscope

Essential components of a typical SEM are shown in Fig. 8.1 The source of electrons is a typical tungsten filament or LaB6 electron gun and is replaced by a field emission gun in modern microscopes.

A Schottky type of field emission gun, which is thermal type, is mostly employed in the SEMs. The filament is heated and also subjected to a high electric field at the same time to extract electrons as described in Chap. 4. It gives a beam of high current density and has a very narrow spread in energy. The electron beam, however, needs further focussing to form a fine spot by a condenser lens. Since the accelerating voltages used for beam acceleration are of a lower range in SEM, 1–30 kV, when compared to a TEM, the condenser lens system (two lenses) is less strong. A set of four scan coils (two pairs) are located just below the condenser lens system which rasters the electron beam synchronously with the electron beam on the display monitor. We will see in a later section that when the electron beam strikes the surface of the specimen a large variety of signals is generated by it from a volume that is much larger than the cylindrical volume under the footprint of the electron beam.

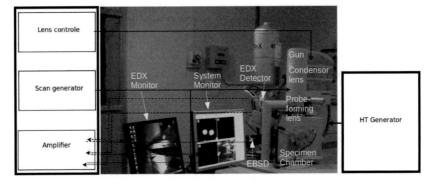

Fig. 8.1 Quanta 200FEG SEM, FEI make along with some ancillary equipment shown in schematic mode on either side

A scanning probe, instead of a stationary one, is essential for this as well as for another most important reason that a lens is not collecting all the signals (even of one type) at the same time in emanating from the region under observation to form an image in this case. This reason becomes more clear when we discuss detectors in a forthcoming section. A final lens, which is often, though inappropriately called objective lens due to its proximity to the object (specimen), acts as a condenser and focusses a fine beam of electrons on the specimen surface. The design of this lens is also different in different microscopes, some of which incorporate certain types of signal detectors within this lens assembly.

In recent models, the specimen stage is of turret type which can accommodate seven specimens, including the central one. The specimens are mounted on small round aluminium stubs. Its distance from the probe-forming lens being adjustable, gives room for positioning different types of detectors and cameras. Many applications of SEM derive benefit from the availability of a large room for mounting different specimen stages such as a straining stage for in situ deformation of specimens and observation of deformation and fracture dynamics, stages with electrical feedthrough for the observation of functioning of integrated circuits. Cryo as well as heating stages are also available for specific experiments.

A scintillation type of detector known as the Everhart-Thornley detector is commonly used to detect the signal electrons. Its shown schematically in Fig. 8.2. It consists of a collector grid and a screen located at an appropriate angle to receive both SE and BSE signals with a bias voltage. The collected signals reach a scintillator which consists of a light pipe that is coated with a metal and kept at +10 kV potential.

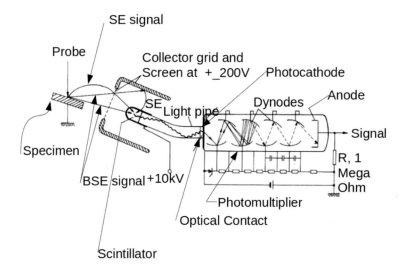

Fig. 8.2 Schematic diagram of the Everhart-Thornley detector [after L. Reimer, Scanning Electron Microscopy. (Springer, Berlin) 1985 as quoted in "Science of Microscopy" eds. Peter W. Hawkes and John C. H. Spence, Vol. 1, Springer Science+Business Media, LLC, 2007. With permission]

An electron that enters the scintillator emits or scintillates a photon carried along the light pipe to a photocathode. The photocathode generates electrons plurally with the help of dynodes which reach the anode and form the signal.

Presently, different types of detectors for different modes of operation are available, which are presented in a later section. A detector appropriate for the chosen signal from the specimen collects the signal which is processed and amplified in a multichannel analyser to be fed to the monitor for rendering the image. No doubt, the scans are synchronised at both ends. In the case of present-day digital monitors, the images are transferred frame by frame.

It is not sufficient to form an image by collecting the signals from specimen maintaining a point-to-point correspondence between the specimen and image since we need magnification to be able to view the image. Increasing magnifications are achieved by scanning ever-decreasing areas (diagonal length say d_1) on the specimen since on the image side the screen is of fixed dimensions (diagonal length say d_2). The magnification at any stage is given by $\dfrac{d_2}{d_1}$.

8.1.2 Types of Signals Generated by the Specimen

A 30 keV electron beam penetrates to a depth of about a micron in the specimen. The depth of penetration is dependent on the atomic number of the specimen material. The probe electrons are scattered by the atoms of the specimen material either in elastic or inelastic mode. Since a majority of the applications use electrons that are scattered inelastically and elastically for image formation, we will first consider these two scattering phenomena.

8.1.2.1 SE and BSE Electrons

Inelastic scattering takes place when the probe electrons lose part of their energy to the valence or conduction band electrons of the specimen material and suffer small energy loss, which is up to 5 eV. In the process, they also suffer a few milliradians of angular deflection. As a result of energy transfer, plasmon excitations and inter- and intra-band transfers may also take place. The typical energies are up to 50 eV. Further, inelastic scattering may also take place with the core-shell electrons. The important point to be understood is that inelastically scattered electrons can come out not only from the atoms at the surface but also from atoms located a few nanometers in depth (usually 5 nm). The probe electrons have sufficient kinetic energy to penetrate further into the depth of the specimen and undergo elastic scattering events many times. As the scattering is elastic, the momentum is conserved by a change of direction. we can expect, then, the elastically scattered electrons, termed Backscattered Electrons (BSE), to come out of the specimen surface at different angles (0–180° in reality). Some of the BSE electrons can also undergo inelastic collisions as they approach

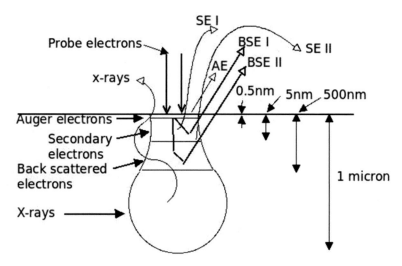

Fig. 8.3 Signals from specimen surface. Note that the indicated depths are the maximum values from which the particular signal can escape the surface. They do get generated at shallower depths. The typical values indicated are also material dependent, such as a metal or non-metal, etc.

the surface. Such inelastic collisions give rise to secondary electrons known as SE II. Those BSE electrons which result from elastic scattering in multiple stages in the spread volume of interaction are termed BSE II electrons. Elastically scattered probe electrons which arise from regions (small area) right beneath the point of incidence of the probe are termed BSE I electrons. The depths, from which the SE and BSE electrons as well as the Auger electrons and X-rays, are indicative of their escape depths. In reality, all these types of signals are generated at all the depths of penetration of the probe electrons, but only those that have sufficient energy to escape to the surface are generated at the indicated depths.

In both the above cases, there is no strict spatial correlation between the signal electrons and the location in the specimen from where they arise. This is due to the spread of the interaction volume under the probe (Refer to Fig. 8.3).

8.1.2.2   Auger Electrons

When BSE electrons knockout a core-shell electron from an atom of the specimen, the particular atom returns to its ground state by the filling of the core-shell vacancy by an electron of its outer shell. In this process, an electron of energy E_A is emitted by the particular atom, which is the difference of energy between the two shells involved in the process of transition less the ionisation energy of the outer shell from where the electron transfer took place. The emitted outer shell electron is known as Auger electron and its energy is in the range of a few keV, albeit much smaller than the energies of elastically scattered BSE electrons and much higher compared to SE electrons. Hence, these are easily detectable in principle. Nevertheless, they often get masked because of inelastic collisions that BSE electrons might have undergone close to the surface that led to the generation of SE II electrons.

8.1.2.3 Photons

Photons of different energies or wavelengths, which are either direct X-rays and fluorescent X-rays or visible light in the case of semiconducting and insulating materials, are also emitted by the specimen. Almost invariably, the X-ray signals are collected to know the local distribution of constituent elements of the specimen to compliment the information gained from the image formed by other signals.

X-rays: As the probe enters the lower depths of the interaction volume shown in Fig. 8.3, X-ray photons get generated since the probe electrons set the dipole of the atoms into oscillation. It is a continuous spectrum of X-rays that is emitted as a consequence. Probe electrons are also strong enough to knock out core electrons of the atoms from their K, L, M....shells. When this happens an electron from the upper shell of the atom jumps to the vacant energy state releasing the difference in energy ΔE between the two concerned shells as an X-ray with a characteristic energy or wavelength. For example, if the jump takes place from the L shell of the atom to the vacancy in the K shell, the emitted characteristic X-ray is designated as K_α radiation in general. From which of the sub-shells of L that the transition takes place decides the designation whether it is $K_{\alpha1}$ (L_{III} to K) or $K_{\alpha2}$ (L_{II} to K), the transition from L_I being forbidden. Similarly, K_β characteristic radiations are due to jumps from M shell to K shell. The white spectrum begins from a minimum energy and spans a wide range of wavelengths or energies with a peak in between. The position of peak is dependent on the energy of incident electrons, while the characteristic peaks occur at fixed energy values that are material-dependent. It is this feature which helps in identifying the elements present in the specimen material. The continuum X-rays, also known as white radiation, are not useful for this purpose. A portion of the characteristic X-rays may impart their energy to some other atoms and generate fluorescent X-rays.

Cathodoluminescence: In the case of semiconductors and insulators, the probe electrons excite electrons from the filled valence band into their empty conduction band, creating electron-hole pairs. Recombination of the electron-hole pairs can take place either directly or indirectly through intrinsic structural defects or extrinsic traps due to impurities. Often non-radiative jumps take place when the excited electron from the conduction band directly recombines with the hole in the valence band. After the recombination, the crystal returns to the ground state. In the case of insulating materials, when the jump takes place from the temporary trapped state, the difference in energy is emitted as a photon with wavelength that is usually in the visible range. It can also be having wavelength anywhere between IR and UV. There is a difference between fluorescent and phosphorescent emissions. The former lasts only up to 10^{-8} s after the stoppage of incoming radiation, while the latter persists beyond

10^{-8} s after the incoming radiation stops. This phenomenon depends much on the band structure of the specimen material.

In semiconducting materials again, the direct recombination of electron-hole pairs emits phonons. It can also occur via the indirect path of extrinsic (impurities or dopants) and intrinsic (defects such as vacancies or non-stoichiometries) recombinations.

Other types of signals such as EBIC and Accoustic are also available for imaging with the use of appropriate detectors. EBIC signal is the Electron Beam-Induced Current (EBIC) that occurs when the electron beam strikes a semiconducting specimen. The current is in terms of electron-hole pairs and can be measured when ohmic contacts are made, one for the electrons and the other for holes. Generally, the currents are of the order of microamperes.

Accoustic waves are also generated when the probe electrons of 20–30 keV strike the specimen which form the basis for a scanning accoustic microscopy.

8.1.3 Principal Modes of Operation

Most applications of SEM use two principal signals, viz., SE and BSE, for image formation. There are new advancements in instrumentation due to which other modes have become available and we will explore these in later sections. The parameters on which the yield of SE and BSE signals depend need to be understood first.

8.1.3.1 Yield

Secondary electrons are those emitted with energy reaching a maximum of 50eV, although a predominant number of them have an average energy of 1–5 eV. Their yield, n_{SE} is more in the case of insulators in comparison to metals, as schematically shown in Fig. 8.4.

Electrons with energies greater than 50eV are termed BSE electrons. These result mainly from elastic scattering events as mentioned earlier and hence show up as a small peak at E_o, the incident electron energy. Many of them participate in both elastic and inelastic events and hence their energy peaks at a value lesser than E_o. Their yield, n_{BSE}, is nearly half of n_{SE} for insulators and is almost equal for conductors. Auger electrons show up as a small peak in the lower range of energies of the BSE regime since their yield is very small.

8.1.3.2 Secondary Electrons

Escape depth: Since their energy is very small, they can only escape to the surface of the specimen from depths as small as 5 nm. Escape depth is also dependent on the material of the specimen, each chemical constituent of which has a characteristic

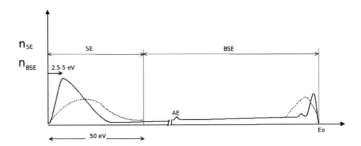

Fig. 8.4 Yields n_{SE} and n_{BSE} as a function of their energy. Solid line corresponds to insulators and dotted line corresponds to conductors

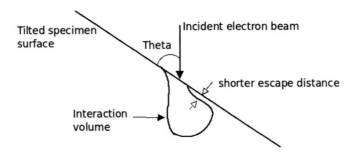

Fig. 8.5 Effect of tilting the specimen on escape distance

mean free path for electrons. We can decrease the effective depth by tilting the specimen surface towards the incident beam as shown in Fig. 8.5.

Essentially, the interaction volume is brought closer to the surface by this tilt and SE electrons generated at greater depths would find a shorter path to the surface.

Effect of E_o: The coefficient of yield, n_{SE} increases with the increase in accelerating voltage or energy, E_o, reaches a maximum and then decreases. The difference in yield between metals and insulators is illustrated in Fig. 8.4.

Effect of atomic number: The yield has no influence on the atomic number of the constituent elements of the specimen.

SE II electrons: These are emitted as a consequence of inelastic scattering of BSE II electrons, while the latter make their way to the surface of the specimen. This event takes place away from the direct impact area of the probe electrons. They are also collected by the same PMT detector and contribute to the image. In case an in-lens detector is used to collect the SE electrons that spiral around the incident beam, the SE II electrons can be stopped from collection and contribution to image formation, as they are emitted away from the impact area of the beam electrons.

8.1.3.3 BSE Electrons

Energy and Escape depth: Since these electrons are the elastically scattered part of the incident beam electrons, they should in principle have the same energy as E_o. A very small peak does show up there as mentioned earlier, but majority of the BSE electrons have a peak energy less than this due to the loss already suffered in inelastic collisions. Their escape depth increases as E_o increases, i.e. electrons that are elastically scattered at greater depths inside the interaction volume can reach the surface.

Depth of penetration: In fact, it is the inverse of the escape depth. In case of true elastic scattering, the two parameters should be equal. Penetration depends on the atomic number of the material. Primary and secondary (BSE II outside of the footprint of the probe) elastic events take place up to a depth of 50 nm.

Yield: It strongly depends on the atomic number and increases non-linearly and monotonically with increasing atomic number (weighted average, in case the material is a mixture of elements) of the specimen material. For the low atomic number elements, the yield is more at lower E_o value (<5 keV) and decreases as E_o. This strong dependency on the atomic number in fact helps in obtaining 'Chemical Contrast' in the image.

Angle of incidence: The yield, n_{BSE} is found to maximise when the beam is incident at about 70–80° angle to the plane of the surface.

8.1.4 Resolution

Since image is not formed by a lens, the concept of resolution and limit of resolution have to be understood in a different way. Depending on the chosen magnification, the area of the specimen surface to be scanned is pre-decided and also the number of scan lines in it. When the probe starts scanning it moves from point to point as shown in Fig. 8.6 At each point, a plethora of signals emerge that are collected by a detector by making a choice from them. Then next point of incidence of the beam should not be too far from the earlier one nor should it overlap, for optimum resolution. The point of incidence itself is debatable—Is it the diameter of the beam or the interaction volume beneath it? Obviously, the interaction volume is much larger compared to the volume of the beam penetrating the depth and also is larger for low atomic number elements and smaller for the high atomic number elements. Only in-lens detector in FEG SEMs (FESEMs) can sample the inner spot in the diagram (corresponding to the beam penetrating volume) (Fig. 8.6) which thus gives rise to a higher resolution. The resolution, therefore, also becomes dependent on the accelerating voltage being used, the mode of image formation and also the mode of operation of the microscope (discussed in Sect. 8.1.7) The corresponding values are illustrated with respect to Quanta 200FEG microscope in the following:

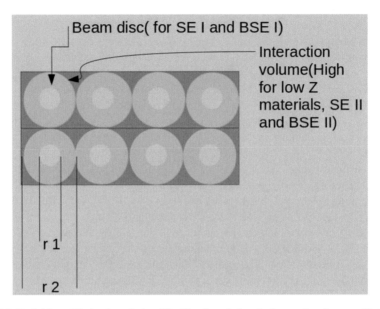

Fig. 8.6 Definition of limit of resolution (If white discs define the beam, then they would also be closely touching each other)

High vacuum	1.2 nm	at 30 kV (SE)
	2.5 nm	at 30 kV (BSE)
	3.0 nm	at 1 kV (SE)
Beam deceleration	3.0 nm	at 1 kV (BD mode + BSED)
(LVSEM mode)	2.3 nm	at 1 kV (BD mode + ICD*)
	3.1 nm	at 200 V(BD mode + ICD)
Low vacuum	1.5 nm	at 30 kV (SE)
	2.5 nm	at 30 kV (BSE)
	3.0 nm	at 3 kV (SE)

8.1.5 Contrast

Signals collected by a detector are prone to some noise due to various factors external to the detector in addition to the internal noise. If we ignore the noise or consider an ideal situation, contrast in the SEM image arises due to the fluctuations or variation in the Signal S, as the beam moves from one point to another. If an average signal strength, S_{av} can be estimated for all points, then Contrast C can be defined as

$$C = \frac{(S - S_{av})}{S}$$

No contrast arises in the image if $S=S_{av}$ and it becomes perceptible if C is at least 5×10^{-5}. Contrast depends on several factors such as topography, chemical constitution, crystal orientation and conductivity of the material. It can also be artificially enhanced by different coatings.

8.1.5.1 Topographical Contrast

It is highest in the case of SE image due to a light-and-shade effect provided the specimen is topologically variable. This is shown in Fig. 8.7 in which case the SE detector is to one side of the specimen. In the case of a biological specimen or an insulator, shadowing is done by coating the specimen with a high atomic number

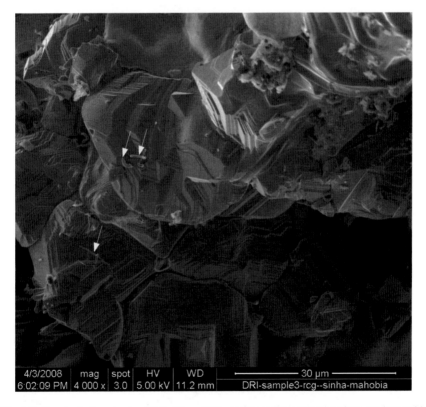

| 4/3/2008 | mag | spot | HV | WD | 30 μm |
| 6:02:09 PM | 4 000 x | 3.0 | 5.00 kV | 11.2 mm | DRI-sample3-rcg--sinha-mahobia |

Fig. 8.7 Topological contrast exhibited by a DRI specimen: The stepped regions are iron oxide that is being reduced in the process and the white arrows point to the whiskers of iron that resulted. [Micrograph reproduced through the courtesy of R. C. Gupta, O. P. Sinha and G. S. Mahobia, IIT (BHU), Varanasi, India, 2008, unpublished work]

metal such as gold or platinum from one side so that a hill-and-valley effect is created due to a higher yield of SE electrons from regions where the deposit is present.

Excellent topological contrast is obtained in the case of a directly reduced iron ore given in Fig. 8.7. It shows the crystalline facets of iron oxide and small whiskers of reduced iron (marked by white arrows in the figure). The specimen was imaged intentionally at a low voltage (5 kV) to minimise charge build-up on the oxide surface.

When the specimen is flat as in the case of a micron thick slab of a quasicrystalline alloy grown by molecular beam epitaxy technique, the topological contrast is minimal and it is medium in the case of the fractured insulator substrate as presented in Fig. 8.8.

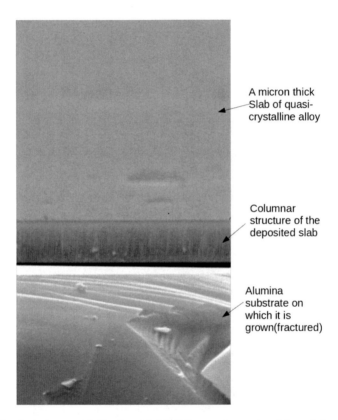

A micron thick Slab of quasi-crystalline alloy

Columnar structure of the deposited slab

Alumina substrate on which it is grown(fractured)

Fig. 8.8 Topological contrast in quasicrystal slab [Micrograph reproduced through the courtesy of Dr. Vincent Fournee, School of Mines, Nancy, France. Specimen grown as a part of an Indo-French collaborative Project No. 33308-1]

8.1.5.2 Atomic Number Contrast

It is also referred to as chemical contrast and arises in the case of BSE images owing to the latter's strong dependence on the elastic scattering strength of the specimen material. The lower atomic number elements scatter less and hence appear dark, while the higher atomic number elements such as Au, Pt, etc., elastically scatter more electrons and hence appear bright in the BSE image. Figure 8.9 shows a BSE image of a nickel-iron base superalloy that is given a thermal barrier coating of 8%Yittria-stabilised Zirconia. The image is taken from the cross section to inspect the adherence of the coating. The high atomic number of the base metal displays it in bright contrast, while the coating in dark contrast.

This contrast helps in rudimentary chemical analysis on the basis of which, phase identification in a multi-component alloy system can be performed. Figure 8.10 illustrates such a case where a quaternary alloy of composition $Al_{62.2\pm0.1}Pd_{20.4\pm0.5}Mn_{8.9\pm0.1}Ga_{8.5\pm0.1}$, selected from a series of alloys with varying Ga content $Al_{(70-x)}Ga_xPd_{20}Mn_{10}$, shows segregation of chemical constituents which stabilise a desired quasicrystalline phase, $Al_{63}Pd_{21}Mn_8Ga_{10}$, into a particular microstructural feature (marked as 3 in the figure). Here, three microconstituents show distinctly bright, grey and dark contrast. The bright contrast phase (marked as 1) is obviously rich in Mn and Pd metals as per the atomic number contrast displayed by BSE images. By X-ray diffraction analysis the composition of this phase is found to be $Al_{58}Pd_{37}Mn_1Ga_4$. In a similar way, the dark phase is identified as an intermetallic T-phase having a composition $Al_{60}Pd_{12}Mn_{17}Ga_{11}$. If Pd and Mn are considered necessary for the stabilisation of the quasicrystalline phase with optimum Ga content in the quaternary Al-Ga-Mn-Pd system, then we can compute the average atomic number per atom of the pair and compare the contrasts. However, the complex phase equilibria in the quaternary alloy may give rise to intermetallics which are rich in these two and

Fig. 8.9 Atomic number Contrast in BSE image [Micrograph reproduced through the courtesy of Ms. K. V. Radhika; Project on Scanning Electron Microscopy Studies on Inconel-718, PM2000 and Russian Copper, Vikram Sarabhai Space Centre, Thiruvananthapuram, 2008]

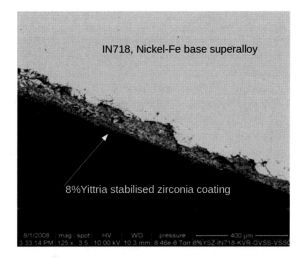

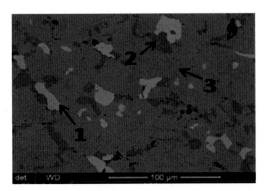

Fig. 8.10 (1.) Inter-metallic phase rich in Mn and Pd(determined by EDX analysis and X-ray diffraction), (2.) is another intermetallic called T-Phase and (3.) the Quasicrystalline phase in the Al-Ga-Mn-Pd alloy. Refer to text for details. [after Murtaza Bohra, M. C. de Weerd, Vincent Fournee, R. K. Mandal, N. K. Mukhopadhyay, Ratnamala Chatterjee and G. V. S. Sastry, Journal of Alloys and Compounds 551 (2013) 274–278. With permission]

consequently give rise to brighter contrast (as 1). Therefore, if we need to identify the quasicrystalline phase among the observed number of phases from the contrast in the BSE image, we should work out the average atomic number per atom for all the expected phases in the system and know the contrast expected from the desired quasicrystalline phase, be it dark, or gray or bright.

8.1.6 Depth of Field

SEM has greater depth of field compared to an optical microscope both being image rendering instruments of the surface of a specimen. This fact is well testified by the SE images in particular as seen in Fig. 8.7. We can arrive at an expression for depth of field in terms of optical parameters of the final probe-forming lens, under certain assumptions. We assume a ray diagram for extremely small spot size, as shown in Fig. 8.11.

Here, R is the radius of the objective aperture and of the lens which focusses the beam to a fine point of dimensions d_1 on the mean surface plane of the specimen. Let ΔU be the depth of the specimen measured in terms of distance from the topmost point to the bottom-most point on the surface of the specimen. Let the electron beam subtend a semi-vertex angle of β on the mean surface. Then the beam spreads to a size of $\frac{d_1}{M}$ on the bottom-most surface.

By the principle of similar triangles

$$\frac{\Delta U}{WD} = \frac{\frac{d_1}{2M}}{R}$$

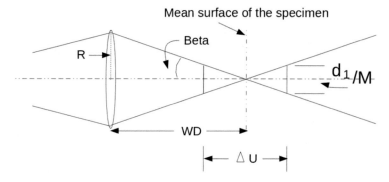

Fig. 8.11 Depth of field

where WD is known as working distance, the distance between the probe forming lens and the surface of the specimen.

Since $\tan\beta = \beta$ for small values of β, we have

$$\Delta U = \frac{d_1}{2M\beta}$$

We can derive an expression on similar lines for a more realistic finite-sized spot with diameter d_2 at the mean surface of the specimen.

8.1.7 Enhancing the Resolution and Other Capabilities of SEM

We know that the resolution is strongly dependent on the spot size created by the probe, keeping in view that there would be some degree of spread of it in the specimen eventually. Then the only way of enhancing the resolution is to decrease the beam diameter and increase the beam coherence to the extent possible. Cold-field emission guns as the electron source are the ultimate in this method at present. The disadvantage being less intensity or current density in the beam on specimen surface. Lower current densities result in weak emitted signals from the specimen and poor $\frac{S}{N}$ ratio. A Schottky field emission gun, which employs some degree of thermal emission in addition to field emission, is usually the choice for SEM. Such models of SEMs are termed FESEMs. A photograph of the Schottky FEG and its functioning is given in Chap. 4. The current density and spot size obtained in, e.g. Quanta 200 FEG SEM using such a gun, are 100 nA (Continuously Variable to this level) and 1 nm, respectively. The resolution achieved in this model is 1 nm in SE mode at 30 kV. Figure 8.12 shows a high-resolution image of multiwall carbon nanotubes (MWCNTs) in SE mode. The large working distances and availability of multiple

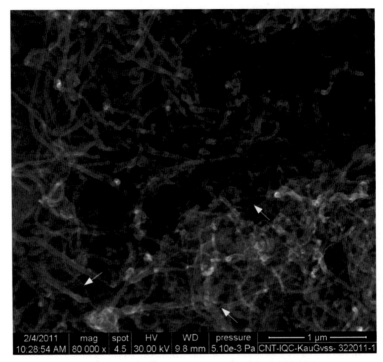

Arrows indicate extremely thin MWCNT

Fig. 8.12 Secondary electron image of Multiwall Carbon Nanotubes (MWCNTs)

detectors in an FESEM not only enhanced the resolving power but also its capabilities to function in low accelerating voltages(LVSEM) and under different specimen environments (ESEM). The Quanta 200 FEG model, for example, has two types of detectors for SE electrons and two types for BSE electrons.

8.1.7.1 LVSEM

With a field emission gun, it is possible to reduce the accelerating voltage and yet obtain a reasonably small spot size with low current density. This feature helps in imaging insulator specimens which otherwise show very low contrast due to charge build-up on their surface. It is also much convenient to use low accelerating voltages in the case of beam-sensitive specimens. In LVSEM mode, the microscope can operate from a very low accelerating voltage of 500 V–about 3 kV, the upper limit being 5 kV. Figure 8.13 shows BSE images of an irradiated quasicrystalline thin film.

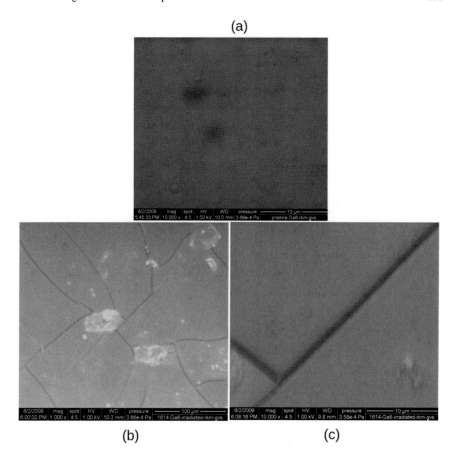

Fig. 8.13 BSE images of an irradiated quasicrystalline thin film in LVSEM mode at 1 keV **a** Pristine thin film of a quasicrystalline alloy in amorphous state after deposition. **b** The same specimen after irradiation. Note the development of crystallinity. **c** Same as **b** at higher magnification, showing grain boundary grooves [Micrograph reproduced through the courtesy of Indo-French centre for the Promotion of Advanced Research, from the unpublished work of Project No.3308-1, 2008]

Since the film is already irradiated and the authors wanted to examine the effects of irradiation without any further dosage from the probe electrons, LVSEM mode was selected to image at 1 keV. The pristine thin film of the quasicrystalline alloy composition resulted in an amorphous state after deposition. After irradiation, the specimen crystallised exhibiting a polycrystalline microstructure with many grains. Note the development of crystallinity. Examination of the boundaries at higher magnification revealed grooving of the grain boundaries. Such features are common in irradiated specimens or specimens subjected to thermal etching. The limits in the operative range of LVSEM arise from the fact that a minimum voltage is required to knock out an electron by inelastic collision and even an elastic scattering requires certain minimum energy which depends on the scattering strength of the chemical constituents of the specimen material. Besides the above, other restrictive factors in the operation of an FESEM in LVSEM mode are the special geometry of the detectors and aberrations of the beam.

The technique offers many more advantages as compared to the difficulties mentioned above, foremost being the ability to observe insulator materials without applying any conductive coatings on them. In LVSEM mode, we get access to SE I and BSE I electrons since the interaction volume gets reduced severely. This gives scope for images with better resolution though less contrast. The yield of SE electrons also is high because of shorter distance to travel to the surface. Therefore, the S-to-N ratio improves. The BSE images no longer show chemical contrast because their yield, n_{BSE} becomes electron energy-dependent. Nevertheless, the chemical contrast and even topological contrast remain feasible for imaging. Figure 8.14 shows a dense covering of the leg of a mosquito with fine feathers. The finer details of the structure of these fine feathers are also resolved in the image which is recorded at 10 keV due to the beam sensitivity of the specimen. With regards to instrumentation, the LVSEM requires a field emission gun as a source of probe electrons, suitably designed detectors and even special lenses such as combined electromagnetic and

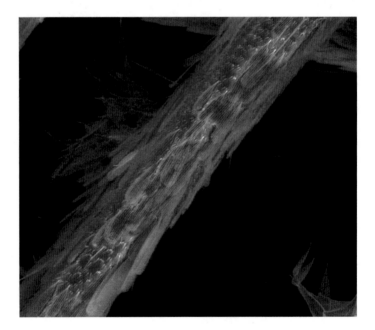

Secondary electron image of a leg of a mosquito,
imaged at 10keV because of the beam sensitivity of the
specimen

Fig. 8.14 LVSEM image: Leg of a mosquito

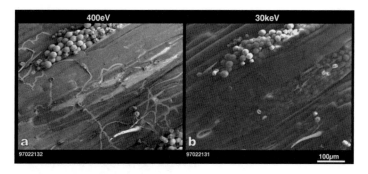

Secondary electron micrograph pair of the cuticula of a leaf recorded at electron energies of 0.4 (a) and 30 keV (b) with an "in-lens" SEM. The low-energy image contains information only from the surface whereas the 30-keV image also reveals information about structural features below the surface, e.g., new spores, which are not visible in (a). (Micrographs kindly provided by Dr. R. Wepf, Beiersdorf AG, Hamburg, Germany.)

Fig. 8.15 Subsurface structures of leaf. A comparison of LVSEM and normal modes [as quoted by R. Reichelt in "Science of Microscopy" eds. Peter W. Hawkes and John C. H. Spence, Vol. 1, Springer Science+Business Media, LLC, 2007. With permission]

electrostatic lenses. We can exploit the capabilities of LVSEM in rendering images of the very top layers of the surface of biological specimens to expose the subsurface structures (see Fig. 8.15). This can be achieved by imaging the same specimen both in LVSEM mode and normal mode.

8.1.7.2 SEM at Elevated Chamber Pressures

Given the versatility of SEM applications ranging from metallic materials, semiconductors, polymers, rocks and minerals, archaeological specimens to biological and forensic studies, it became imperative to preserve the surface of the specimens in their pristine, native condition. In addition, there is a demand for instruments that enable observation of interface structure during gas/metal reactions.

Three essential components of procedure are a differential pressure aperture, a mechanism to retain the moisture in biological specimens and a method by which detectability of the weak signals is enhanced.

Differential pumping is effected by providing a pressure limiting aperture (PLA) between the main column and the specimen chamber (Fig. 8.16). The main column requires a pressure as low as 10^{-4}–10^{-5} Pa in an FESEM, while the specimen cham-

Fig. 8.16 Differential pumping apertures, PLA1 and PLA2 [Photograph reproduced through the courtesy of FEI Company, USA through Icon Analytical Equipment, Pvt. Ltd., India]

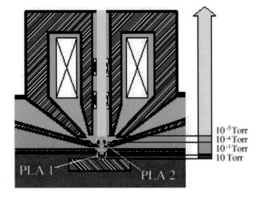

10^{-5}Torr
10^{-4}Torr
10^{-1}Torr
10 Torr

ber and neighbourhood of specimen, in particular, require a pressure of few tens of Pa to few thousands Pa (SE electron imaging).

Environment: An environment can be created either of water vapour or any gas in the specimen chamber depending on whether a biological wet specimen is being studied or the reaction kinetics at the gas/material interface. The surface of a biological material can be retained in its equilibrium state provided the water inside the material is in equilibrium with water vapour in its surrounding environment. We can know the pressure of water vapour required to be maintained in the vicinity of the specimen from the equilibrium diagram of unary system of H_2O. The difficulty with this approach is that even at an ambient temperature of $20°$, the vapour pressure should be as large as 2300Pa. Before, the advent of ESEM researchers used to either freeze-dry the specimens or adopt other specimen preparation techniques that retain the inherent structure to some extent. We can retain sufficient water vapour close to the surface of the wet specimen and yet approach it with the probe electrons. As the beam electrons reach near the surface, they get scattered by the water vapour or gas molecules (whichever may be the environment) as shown in Fig. 8.17. The number of scattered electrons increases as the mean free path of the gas molecules decreases and the distance to the specimen increases. The mean free path in turn is inversely proportional to the pressure of the gas. Thus, a small shroud of electrons gets created around the beam, near the surface of the specimen (represented by black dots in Fig. 8.17). The actual area of the specimen sampled by the beam, thus happens to be the impact area only and the shroud electrons in effect behave as interaction volume as in the normal SEM mode. The emitted SE electrons or BSE electrons travel upwards to the detector. Their initial number is very small because of the smaller scattering strength of atoms of biological specimens and also because some of the beam electrons are scattered away by the gas molecules. As they move towards the detector, they encounter the gas molecules and suffer collisions once again. These collisions are beneficial because they give rise to a cascade effect, generating many more electrons that result in an increased signal strength at the detector surface. Effectively, it is an amplification of the signal. The number of such cascade events taking place before reaching the detector is, of course, a function of gas pressure. We

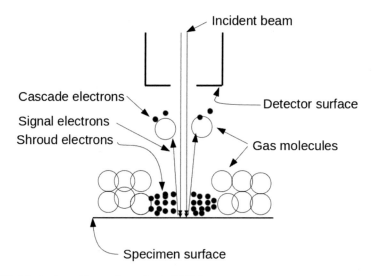

Fig. 8.17 Mechanism of image formation in ESEM

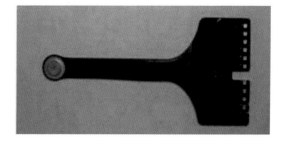

Fig. 8.18 GSED detector used for collecting SE electrons in ESEM

can understand this in relation to the mean free path available to the SE electrons—either too large or too small. We can increase the number of SE electrons reaching the detector by providing a positively charged plate at the detector entry (a GSED detector). The BSE electrons are, on a relative scale, much stronger.

Therefore, we can realise the same resolution as in the conventional FESEM in ESEM mode as well (Refer to Table in Sect. 8.4 for comparison).

Detectors: There are more than one type of detector used in ESEM. The GSED, the gaseous secondary electron detector discussed above provides a better S-to-N ratio for the SE electron imaging and almost excludes BSE electrons due to its geometry (see Fig. 8.18) and its location (clamped on to) on the final probe forming lens. This detector is useful for high-resolution imaging of wet specimens as it can operate under chamber pressures up to 2600Pa. Another type of detector, known as LFD detector, which offers a larger field of view when compared to GSED, is suitable for chamber pressures of a little over 100Pa. The collected signal mostly comprises BSE electrons. It partially surrounds the probe-forming lens cone (see Fig. 8.19) and can easily be clamped into the detector circuit. Its location is such that it can coexist with

Fig. 8.19 LFD detector used for collecting BSE electrons in ESEM

(a)   (b)

Fig. 8.20 LD Converter flue dust, consisting of iron oxide spheres nucleated and grown on graphite flakes

a normal BSE detector of high vacuum mode, while the GSD cannot as it is clamped onto the probe-forming lens.

Applications of ESEM are as diverse as conventional SEM, although it was developed for the need of biological specimens. They range over-examination of tissues, dental implants, metallic or ceramic hip joints, forensic studies such as blood stains, pharmaceutical industries, polymer or natural fibre-based composites (Fig. 8.20).

8.1.8   Electron Backscattered Diffraction

The backscattered electrons are those elastically scattered incident electrons of the probe that come out of the surface in a direction opposite to that of incident electrons. In a specimen that is not optically polished, these BSE electrons, though emerge in straight-line paths, are not collected with any spatial correlation. Hence, they form an image that is not site-specific in a general sense. If the specimen surface were to be optically polished, the crystal planes of a region or even a very small grain would scatter the incident electrons in specific directions that would form a pattern called back scattered diffraction—the forward scattered electrons cannot reach the surface and are absorbed by the specimen. EBSD can give us a precise crystallographic orientation of the small region which is being scanned by the probe at that instant. Thus, crystallographic orientations within a grain of even sub-micron size can be mapped from point to point very precisely. Of course, it requires an elaborate arrangement of detector camera and reference frame for recording and interpreting the EBSD pattern. Such a set-up is shown schematically in Fig. 8.21.

8.1.8.1 Specimen Preparation

As a first step towards acquiring clear, high-resolution patterns, specimens need to be polished with meticulous care. Optical finish is obtained by using the routine metallographic sample preparation procedure adopted for optical microscopy. However, care should be exercised to maintain the bottom and top surfaces of the specimen slab as parallel as possible in this case. In the final step, the top surface is polished to mirror finish using colloidal suspension of alumina. A perfect parallel slab of specimens is mounted on the central hub of a turret specimen holder with either conducting adhesive carbon tape or copper-coated adhesive tape or preferably colloidal silver paste, which should be cured for sufficient time after mounting. Adhesion to the stub is critical as the specimen may slide down in the case of long-duration scans of several hours, or even days. Specifically designed mechanical clamps are employed by some users to do away with the adhesives.

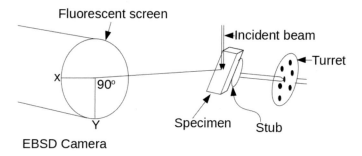

Fig. 8.21 Schematic of EBSD set up

8.1.8.2   Specimen Orientation

The specimen needs to be oriented with respect to the EBSD camera such that the backscattered beam strikes the phosphor screen of the camera at normal incidence. The x- and y-axes on the screen would obviously be along the horizontal and vertical diameters, respectively, of the circular screen as shown in Fig. 8.21. For achieving this orientation, the specimen holder needs to be tilted towards the camera by an angle of $\simeq 70°$ w.r.t. its horizontal position. The tilt required is machine-dependent but usually close to 70°. We should exercise caution while tilting the specimen holder lest the specimen should hit the fragile phosphor screen. Usually, the manufacturer specifies the maximum height of the specimen holder (the working distance) for safe tilting. The IR camera in the specimen chamber can also be of some help. We fix the scan parameters and also obtain an image of the surface a priori and then only tilt the specimen to the acquiring position.

8.1.8.3   Acquiring the EBSD Pattern

An area to be scanned for obtaining the pattern is decided by the specific purpose for which the scan is being carried out. The purpose may be any one of the many applications which are discussed in a later section (Sect. 8.1.8.5). Before fixing up the parameters of the scan, the camera needs to be calibrated. It consists of a phosphor screen behind which a highly sensitive CCD camera is mounted. The pattern glows on the phosphor screen and is recorded by the CCD camera. As the intensity of the raw pattern is very low, sometimes image processing may be carried out but with caution. Therefore, the camera requires to be calibrated w.r.t. the dark current or better known as background subtraction which can be accomplished by adjusting the brightness and contrast of the camera. The number of frames to be recorded and time of exposure, the average of which is finally stored, is to be pre-selected. A large number of frames not only improves the quality of the pattern but also slows down the scan speed. Usually, five frames give good statistics.

A software provided by the vendor acquires the raw patterns from a chosen scan area and processes the data to display requisite vital parameters (discussed in detail in Sect. 8.1.8.4) of the microstructure. The selection of the scan line spacing and the spot size decides the total number of acquired counts. The choice being decided by the R-value of the statistical fit besides being restricted by the inherent resolving power of the microscope and the effective beam diameter. The total time that would be taken for the chosen scan parameters would also be displayed by the software. Depending on this, the time of scan can either be increased or decreased by changing the scan parameters. In another drop-down menu of the software, the crystallographic data of a standard material, i.e. identical to or closest to the material of the specimen is linked through a pre-loaded JCPDS file (for example, we may be dealing with commercial purity Al specimen that is subjected to, say 20% deformation, while the standard material available in JCPDS files may be pure Al).

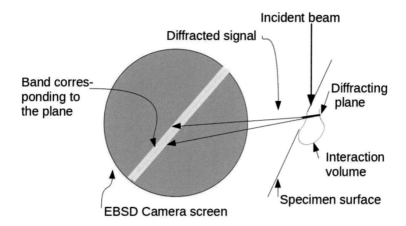

Fig. 8.22 Band detection, schematic

Fig. 8.23 Hugh transform

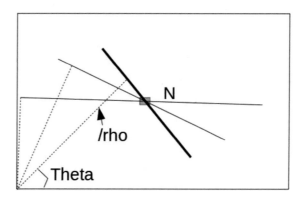

Typically, a pattern acquired in the scan is analysed by the software provided (EDAX-TSL) by identifying different bands formed in the pattern, each being an image of a particular crystallographic plane that is in Bragg condition (see Fig. 8.22). Usually, the patterns are very weak in intensity and are difficult to identify the bands. Therefore, intersection of bands which are known as zones are identified first using an algorithm called Hugh transformation (Fig. 8.23).

$$\rho = rCos\theta + pSin\theta$$

where r and p are the x- and y-coordinates, respectively, of the pixel (column and row) and ρ and θ are the coordinates of the lines that pass through a pixel. Note that there may be many bands that pass through a pixel (or a band may be extended beyond one pixel). Thus, a pixel information is mapped into a sinusoidal curve in Hugh space. As soon as we move to another pixel, n+1th, the r, p as well as ρ, θ values change. Therefore, the data is stored in bins in close intervals of ρ and θ. The

Fig. 8.24 Peaks in the
transform space

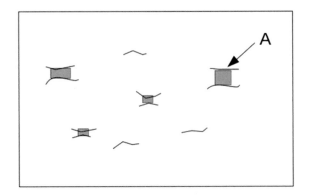

greyscale values depend on the bit configuration, whether it is 8-bit or 16-bit, etc.,
and range from 0 to 255 for 8-bit configuration. The greyscale value of the pixel is
added to each bin along the sinusoidal curve. Thus, each line represented by a set
of sinusoidal curves and a complete EBSD pattern would be like the one shown in
Fig. 8.24.

Note that peaks in the transform space are characterised by two dark bands on
either side of a bright region shown by an arrow at A. After locating such nodes, bands
can be constructed. Each time a pair of bands is taken to identify the crystallographic
planes to which they correspond with the help of JCPDs files mentioned earlier. The
angle between the bands corresponds to the projected angle of a pair of hkl-planes
of the crystal. For a high-symmetry crystal, that step may result in many possible
pairs within specified error limits. Which is the correct possibility, can be guessed
to some extent from the relative width of the bands, since the bandwidth increases
with increasing hkl-indeces. A best way to tackle this problem is to choose a triplet
of bands as shown in the diagram (Fig. 8.25).

Even if this method offers multiple choices for a triplet for a chosen crystal
structure, a voting algorithm is adopted and the highest voted choice is indexed, the
zone axis identified as shown in Fig. 8.26.

8.1.8.4   Microstructural Parameters

Such indexed EBSD patterns at the least, help in identifying the phases in a multi-
phase material. We might take help of EDX analysis of the region in case of any
ambiguity arising out of similar crystal structures but with varying lattice parameters
for the phases being identified.

For most of the applications, the indexed patterns form a valuable background
database. They are used in constructing 2D maps of the changes in orientations of the
microstructure, called Inverse Pole Figure (IPF) maps and texture maps. Pole figures
are a representation in reciprocal space, of predominant orientations of grains of a
polycrystalline material. They are similar to stereographic projections corresponding

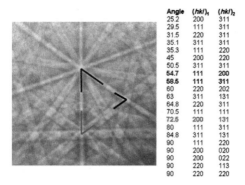

Angle	$(hkl)_1$	$(hkl)_2$
25.2	200	311
29.5	111	311
31.5	220	311
35.1	311	311
35.3	111	220
45	200	220
50.5	311	311
54.7	111	200
58.5	111	311
60	220	202
63	311	131
64.8	220	311
70.5	111	111
72.5	200	131
80	111	311
84.8	311	131
90	111	220
90	200	020
90	200	022
90	220	113
90	220	220

Fig. 8.25 Triplet of bands, a schematic of using a triplet to find a single indexing solution. [Photograph reproduced through the courtesy of EDAX/TSL, as quoted in, B. L. Adams, S. R. Kalidindi, and D. T. Fullwood. Electron Backscatter Diffraction Microscopy and Basic Stereology. Microstructure Sensitive Design for Performance Optimization, 341–371 (2013), Elsevier Inc. With permission]

to all the grains in that orientation but superimposed onto one projection (Fig. 8.27). For centrosymmetric crystals, the possible orientations can be represented by one octant of the projection circle as shown in Fig. 8.28.

If the projection of different crystallographic directions within the octant is taken, they would be circles in extent but would intersect each other at the so-called boundary which may not always be a true grain boundary or subcell boundary. Such a map is referred to as IPF map. The reason for this nomenclature is obvious from the proceeding discussion. The three primary poles of the octant are assigned R, G, and B colours such that all those regions of the specimen whose poles (perpendiculars to the surfaces) are parallel to 001, 110 or 111 are assigned these colours. Any intermediate direction is given the corresponding hue from this palette.

Bulk of the applications of EBSD technique is for constructing such maps which offer solutions to engineering and science problems and also to problems in geology and petrology.

Grain boundaries and sub-boundaries: It is possible to identify and distinguish between high-angle boundaries (grain boundaries) and low-angle boundaries (subcell boundaries) by defining a cut-off angular misorientation value between two regions. Usually, two adjacent grains differing by an angle between 5 and 15° of misorientation are recognised as subgrains or subcells and for angles of misorientation larger than this value, the boundaries are recognised as high-angle boundaries that exist between usual grains. Colour coding of the boundaries makes them easily discernible. These parameters are displayed by the software along with the IPF maps. An example is shown in Fig. 8.29.

Misorientation estimations: We might realise from the above discussion that the crucial step in Orientation Imaging Microscopy (OIM) is the correct representation of intragranular and intergranular misorientations. This parameter can be represented in different ways, each rendering the information in a slightly different perspective. A

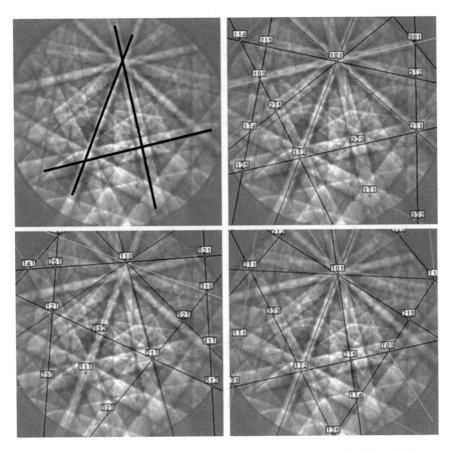

Fig. 8.26 Indexed Pattern, [after B. L. Adams, S. R. Kalidind, and D. T. Fullwood, Electron Backscatter Diffraction Microscopy and Basic Stereology. Microstructure Sensitive Design for Performance Optimization, 341–371, (2013) Elsevier Inc. With permission]

Fig. 8.27 Typical Pole Figures. [after Deepa Verma, Satish Kumar Shekhawat, N. K. Mukhopadhyay, G. V. S. Sastry and R. Manna, Journal of Materials Engineering and Performance ISSN 1059–9495, J. of Materials Engineering and Performance 28 (2019) 3638–3651. With permission]

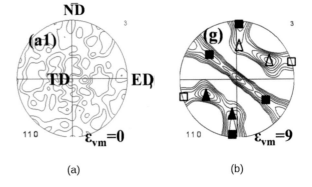

(a)   (b)

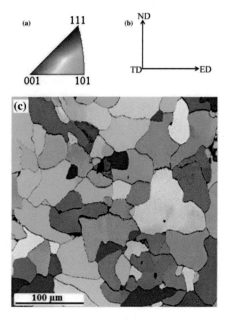

Fig. 8.28 Inverse Pole Figure (IPF) [after Raj Bahadur Singh, Deepa Verma, N. K. Mukhopadhyay, G. V. S. Sastry and R. Manna, Journal of Materials Engineering and Performance, 28(6) (2019) 3638–3651. With permission]

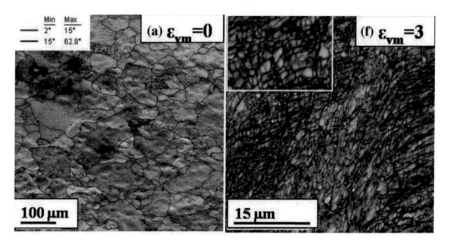

Fig. 8.29 High-angle and low-angle boundaries marked on IPF maps [after Deepa Verma, N. K. Mukhopadhyay, G. V. S. Sastry and R. Manna, Metall. Mater. Trans A, 47A (2016) 1803–1817. With permission]

Fig. 8.30 Definition of
KAM

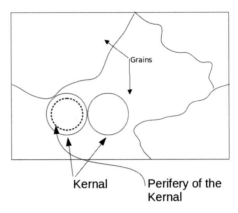

variety of representations is possible from which valuable metallurgical parameters such as strain in the lattice, recrystallisation kinetics, evolution of texture in post-thermomechanical processing condition or upon solidification can be obtained. These are:

1. Grain size versus Area fraction
2. Misorientation angle versus Number fraction
3. Kernel Average Misorientation (KAM) versus Number fraction
4. Misorientation profile versus Distance
5. Grain Orientation Spread (GOS) versus Area fraction
6. Grain Average Misorientation (GAM) versus Number fraction

Of the above, KAM, GOS and GAM are most widely used.

<u>KAM</u> is defined with respect to a user-defined area rather than a grain in totality. Thus, it can be smaller than the actual grain and can be more meaningful as the grain boundary itself is by a definition rather than the physical entity. An average misorientation is defined as an average angle of misorientation between a pair of points inside a kernel with all its neighbouring points along the periphery of the kernels. Kernel average misorientation is an average of all above-mentioned misorientations. For every point, this is defined with respect to its neighbouring points (see Fig. 8.30).

If the neighbouring point happens to fall on a grain boundary, it will give a larger misorientation ($>5°$) and hence is not taken into account while averaging.

<u>GOS</u> is grain-based misorientation estimate. The orientation spread is estimated as the average deviation between the orientation of each point in the grain and the average misorientation for the entire grain which is calculated a priori. The distribution of orientation spread for all the grains in the scan area of the specimen is then represented as GOS.

<u>GAM</u> is estimated by first calculating the angle of misorientation between pairs of neighbouring points and then the average of them is estimated. GAM then represents the frequency distribution of these averages.

8.1.8.5 Application to ECAP Processing

Equal Channel Angular Pressing (ECAP) is a method of accumulating plastic strain in a metal that eventually transforms the microstructure. ECAP processing uses a specially designed die that accumulates severe deformation in each pass of processing, without any change in the dimensions of the specimen. As a result, by a certain number of passes, an extremely high density of dislocations gets stored in the material which forces dynamic recovery and even recrystallisation at the ambient temperature. The signature of recrystallisation that is taking place, can be noticed by monitoring the grain size variations and variations of crystallographic orientations within the grains represented by IPF maps. Besides these parameters, we can also monitor the nature of grain boundaries, i.e. whether they are high-angle type (in recrystallised grains) or low-angle type (subgrains in recovery stage), accumulated strain in the lattice and the like. It is in this respect that EBSD proves to be a powerful tool to monitor the above-mentioned changes and offers a wealth of information.

In the following study, dilute aluminium alloys with different alloying elements that differ in their Stacking Fault Energy (SFE) were subjected to ECAP processing for a different number of passes. The aim of the study was to verify the effect of SFE of the alloy on the number of passes (which is an indicator of the extent of strain accumulation) required to achieve an ultra-fine-grained (UFG)microstructure.

For the sake of brevity, we will demonstrate the power of EBSD analysis by considering only two compositions of Al-Mg alloy for monitoring the microstructural changes taking place after each pass till the UFG microstructure is formed.

The initial microstructure of Al-1Mg alloy is coarse (90–100 μm)as shown in Fig. 8.31a. The corresponding IPF map shows that the microstructure is very random in orientations (b), with 50% of the grains having 80–90 μm diameter and (c) a KAM of only 0.5°, i.e. to indicate that the grains are free of any intragranular strain. After subjecting the specimen to four passes of ECAP, we notice a drastic change in the microstructure as given in Fig. 8.32, correspondingly the grain size estimates (b) and KAM estimates also show great variations. The grain sizes now spread over a range of 1–10 μm is an indication of progressive reduction that is taking place in grain size. The KAM estimates indicate a spread, though of a minor fraction.

By eight passes, the microstructure appears to have reached a final state of refinement with a uniform spread of orientations as indicated by the IPF map represented in Fig. 8.33. The average grain size is 0.4 μm and a KAM with a negligible spread in orientations, i.e.the recrystallised grains are devoid of any internal strain.

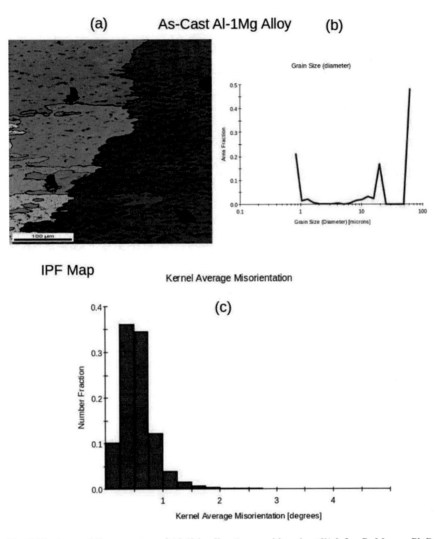

Fig. 8.31 As-cast Microstructure of Al-1Mg alloy (compositions in at%) [after R. Manna, Ph.D. Thesis, IIT (BHU), Varanasi, India, 2008. With permission]

By an optimum amount of Mg addition, i.e. Al-5Mg alloy, the number of ECAP passes required to achieve the desired degree of refinement could be brought down as illustrated in Figs. 8.34, 8.35 and 8.36. The as-cast microstructure has dendrites in it with an average grain size around 90–100 μm and KAM spread over 0.5°.

By two passes, the IPF map (Fig. 8.35) shows considerable refinement to be taking place. This evidence is also supported by the observed decrease of average grain size to 6 μm and a KAM spread up to 3°.

Al-1Mg Alloy, ECAPed for 4 Passes

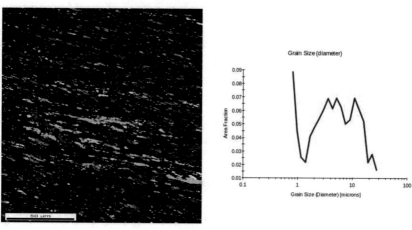

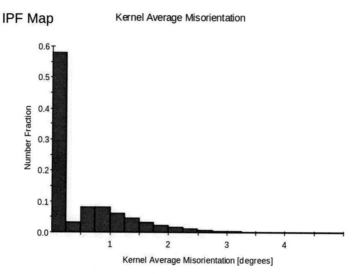

Fig. 8.32 Al-1Mg Alloy ECAPed for four Passes [after R. Manna, Ph.D. Thesis, IIT (BHU), Varanasi, India, 2008. With permission]

Remarkable change can be noticed in the IPF map by Pass 3, which indicates a uniform spread of orientations. The average grain size drops down to about 2 μm with a majority below a micron in size. The KAM spread is negligible. This study added ample justification to the hypothesis (Manna (2008)) that SFE of the chosen aluminium alloy has a significant effect on grain refinement and dynamic recrystallisation achieved through ECAP processing.

Al-1Mg Alloy ECAPed for 8 Passes

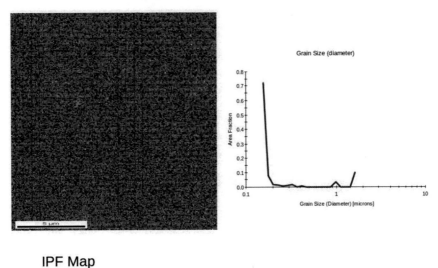

IPF Map

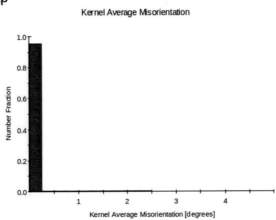

Fig. 8.33 Al-1Mg Alloy ECAPed for eight Passes [after R. Manna, Ph.D. Thesis, IIT (BHU), Varanasi, India, 2008. With permission]

Suggested Reading

1. Scanning Electron Microscopy Physics of Image Formation and Microanalysis, by Ludwig Reimer, Springer Series in Optical Sciences, 1988.
2. Chapter 3, by Rudolf Reichelt in "Science of Microscopy" eds. Peter W. Hawkes and John C. H. Spence, Vol. 1, Springer Science+Business Media, LLC, 2007.

As-Cast Al-5Mg Alloy

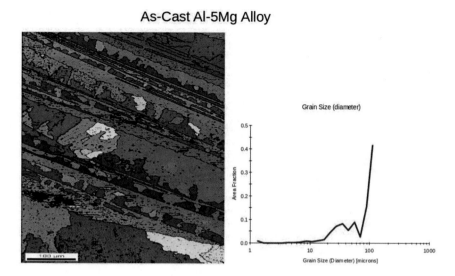

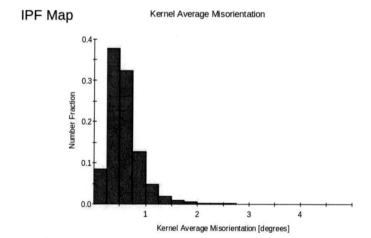

Fig. 8.34 As-cast Microstructure of Al-5Mg Alloy [after R.Manna, Ph.D. Thesis, IIT (BHU), Varanasi, India, 2008. With permission]

3. Electron Backscatter Diffraction Microscopy and Basic Stereology, Chap. 16, in "Microstructure Sensitive Design for Performance Optimization by Brent L. Adams, Surya R. Kalidindi, David T. Fullwood" Butterworth-Heinemann, 2012.

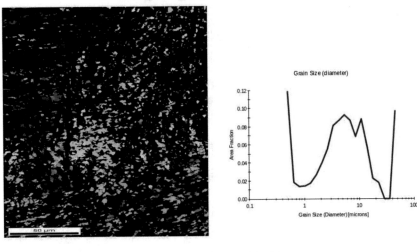

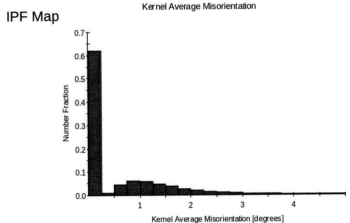

Fig. 8.35 Microstructure of Al-5Mg Alloy [after R. Manna, Ph.D. Thesis, IIT (BHU), Varanasi, India, 2008. With permission]

Exercises

Q1. Why is secondary electron imaging mode preferred over backscattered electron imaging for fracture analysis?

Q2. The depth of field in the case of a scanning electron microscope is observed to be higher than that of an optical microscope. Gives reasons.

Q3. Calculate the depth of field available in a scanning electron microscope, that is operating at 1000x magnification with a spot size of 0.1mm and an objective aperture of 10 μm, for the study of a specimen whose mean surface is at a distance of 10mm.

Al-5Mg Alloy, ECAPed for 3 Passes

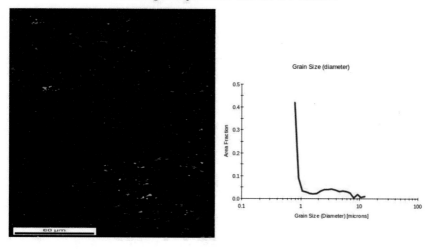

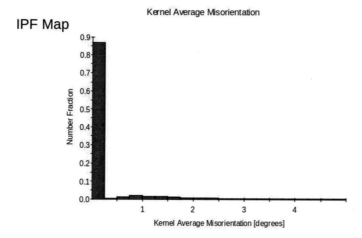

Fig. 8.36 Microstructure of Al-5Mg Alloy ECAPed for three Passes [after R. Manna, Ph.D. Thesis, IIT (BHU), Varanasi, India, 2008. With permission]

Q4. Multiwall carbon nanotubes of μm length are grown on a quasicrystalline substrate. If a specimen of it needs to be observed in an SEM such that both the substrate and the MWCNT coat on it are in focus simultaneously, calculate the working distance to be chosen in the microscope. The collection angle of the secondary electron detector is 0.16 rad, spot size 0.1mm and the magnification 50,000x.

Appendix A
Stereographic Projection

Stereographic projection is used to represent the symmetry elements of a crystal. Indeed visual representation is preferred choice of human mind out of the two available choices, viz., analytical method (trigonometrical formulae or group theoretical matrices) and graphical method (also called crystal projection). In graphical methods again there are four different ways of crystal projections. These are cyclographic, orthographic, gnomonic and stereographic projections.

The stereographic projection itself is based on spherical projection, i.e. projection of a crystal on to a transparent (or even a wire model) sphere at the centre of which the crystal is placed. We will discuss the stereographic projection in detail and refer the readers interested in cyclo, ortho and gnomonic projections to a text book of mineralogy by Dana and Ford (1959).

As a first step, spherical projection is used to construct the stereographic projection. A tiny crystal is assumed to be located at the centre of a transparent sphere. The crystal planes of interest are extended to intersect the spherical surface in great circles–circles on the sphere whose diameters are same as that of the sphere. The angle between the two great circles is the true angle between the two chosen crystal planes (Fig. A.1).

While the spherical projection truly represents the crystallographic information, it is cumbersome to carry the projection sphere around for studying any crystal symmetry. Instead, the stereographic projection method can be used which gives a two-dimensional map of the three-dimensional crystal symmetry information. Though in some of the older texts (e.g. Barret and Massalski 1966) gnomonic projection is referred to as stereographic projection, we will adopt the standard method of stereographic projection.

Crystal planes are represented by their plane normals which intersect the sphere of projection in points, called poles, in stereographic projection. Crystallographic directions are represented by the great circles which are intersections of the planes normal to the corresponding directions, on the sphere. Thus, there is a built-in reciprocal relation between crystal planes and directions (orthogonal cases). The special features of triclinic crystals will be discussed separately.

G. V. S. Sastry, *Microstructural Characterisation Techniques*, Indian Institute of Metals Series, https://doi.org/10.1007/978-981-19-3509-1

Fig. A.1 Angle between
two planes is the angle
between the two
corresponding great circles

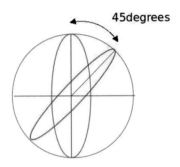

45degrees

A.1   Projection Method

We place the tiny crystal at the centre of the sphere as in the case of spherical projection. The crystal is taken to be of cubic system for simplicity (e.g. a [uvw] direction is perpendicular to the plane (hkl) where uvw and hkl are identical integers) though the method is applicable to all crystals. Considering the sphere of projection as globe, the equatorial plane is considered as plane of projection and the [001] direction of the crystal is coincided with the N-S direction of the sphere as shown in Fig. A.2. The desired poles and great circles are then joined by lines to the south pole. The points and arcs (great circles on sphere project as circles or arcs of true circles on to the equatorial plane.) of intersection of these connecting lines with the equatorial plane is considered to be the stereographic projections of the respective crystal planes and crystal directions (shown in Fig. A.2).

The equatorial circle itself when joined to the south pole intersects the sphere as a horizontal great circle and is termed as Basic Circle. We can easily realise that it is a projection of the [001] direction of the cubic crystal and the (001) plane gets projected at the centre of the stereogram. We would also realise that any pole lying on the southern hemisphere when connected to the south pole would intersect the equatorial plane outside the Basic Circle. In case it is essential, the requisite pole in the southern hemisphere may be shown by the projection of the diametrically opposite one in the upper hemisphere by a dotted circle. It is obvious that all the other poles and great circles of the upper hemisphere project within the Basic Circle. If we choose to project every crystal plane and direction, then the Basic Circle gets densely filled. Therefore, poles of low index planes are projected normally on to a desired crystal plane (hkl) and it is called standard (hkl) stereogram of the cubic crystal in that standard orientation [hkl]. This is true for cubic crystals as planes and plane normals have the same indices in cubic system. This is valid for other crystals as well in some special cases.

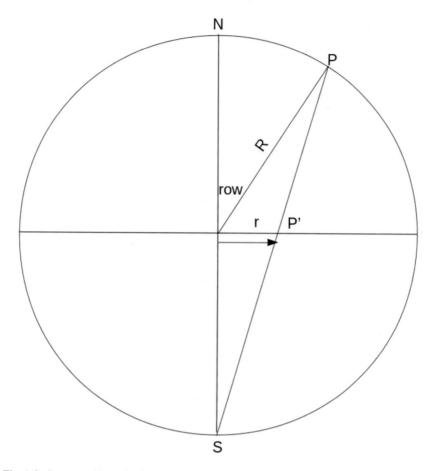

Fig. A.2 Stereographic projection method

A.1.1 *Important Properties*

1. Stereographic projection preserves the symmetry properties of the axis perpendicular to the plane of projection.
2. Circles or arcs of circles on spherical projection are projected as true circles or arcs of true circles.
3. Latitude circles are not great circles on projection, except the equatorial circle. All longitudinal circles are great circles.
4. All crystallographic planes whose poles are parallel to the axis of projection get projected at the same centre point of the Basic Circle.
5. Angular relations between poles are preserved.
6. Angular relations between poles are not changed by rotation of the poles about the axis of projection.

7. The angle between two poles is the difference in their latitudes, when so rotated to lie on the same meridian line.
8. Stereographic projection can represent internal symmetry as well as external features of a crystal.

Essentially we need to measure the angles between poles which can be done by either a protractor or a Wulff net. Standard Wulff nets are available with chosen diameters on which meridian lines and latitude circles are marked at 2° intervals. Such a net is shown in Fig. A.3.

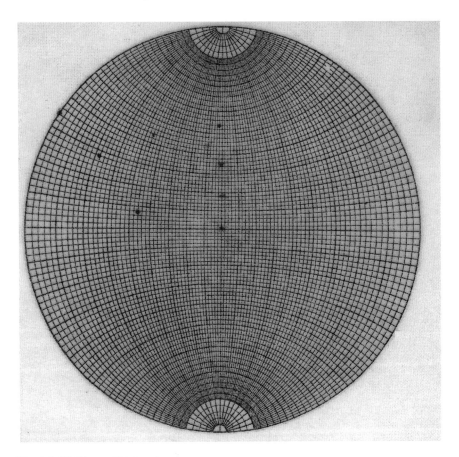

Fig. A.3 Wulffnet at 2° intervals

A.1.2 Construction of Wulff Net with Desired Diameter and Intervals of Lines

We can also construct a Wulff net with desired diameter and interval of lines with the help of analytical equations.

Meridian lines: The projection of pole P in Fig. A.2 is given by

$$r = R \tan\left(\frac{\rho}{2}\right)$$

where R is the radius of the Wulff net and r is the radial distance of projection point P' on the Basic Circle from the centre O. A meridian line can be drawn with radius r_g according to

$$R_g^2 = R^2 + (OP' + P'C)^2$$

where C is the centre of arc of the meridian line.

$$i.e. = R^2 + \left(R_g - R\tan\left(\frac{\rho}{2}\right)\right)^2$$

Therefore,

$$R_g = \frac{R}{Sin\rho} \tag{A.1}$$

Latitude Circles: Let P be a pole as shown in Fig. A.2.

$$OD = R\tan\left(\frac{(90 - \phi)}{2}\right)$$

$$OD' = RCos\phi$$

$$DD' = RCos\phi - R\tan\left(\frac{(90 - \phi)}{2}\right)$$

$$PD' = RSin\phi$$

From the $\Delta C'PD'$

$$R_s^2 = [PD']^2 + [C'D']^2$$

i.e.

$$R_s^2 = \left[R_s - RCos\phi + R\tan\left(\frac{(90 - \phi)}{2}\right)\right]^2 + R^2Sin^2\phi$$

Therefore,

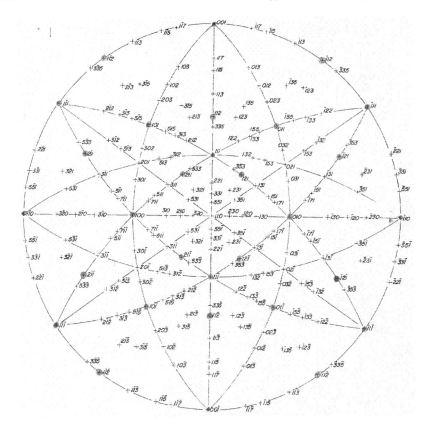

Fig. A.4 Standard cubic [110] projection

$$R_s = R \tan\phi$$

A simple application to measure the angle between two poles has been illustrated in Sect. A.1.2. Readers are suggested to practice the exercises given by Zohari and Thomas in their book (Zohari and Thomas 1969). Standard cubic projections along [001], [110] and [111] are represented in Figs. A.3, A.4 and A.5. These are most commonly required stereographic projections and are also helpful in understanding the indexing method described in Sects. 5.7 and 5.5.1.

A.1.2.1 Non-orthogonal crystal systems

A triclinic crystal has only a centre of symmetry and the crystal axes are inclined to each other at angles other than 90°. Therefore, when we place a tiny triclinic crystal at the centre of projection sphere, the crystal axes are arbitrarily oriented with respect to the Euclidean axes X, Y, Z. Consequently, the poles, when connected to south

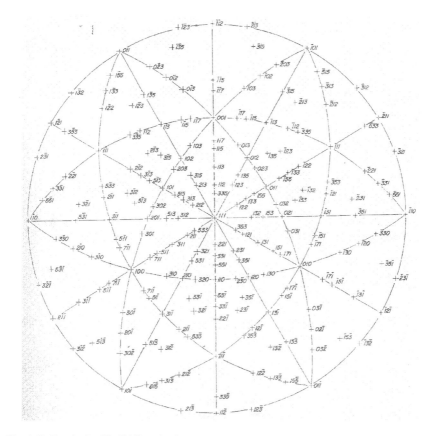

Fig. A.5 Standard cubic [111] projection

pole, do not intersect the equatorial plane at the perimeter of the Basic Circle as do the orthogonal systems. The resulting stereogram is shown in Fig. A.6. If, as in the standard practice, the c-axis of the crystal is oriented along the N-S direction of the sphere, it would project at the centre. The open squares are the projections of the axes on the lower hemisphere.

A.1.3   Applications

Earliest applications of stereographic projection are in the field of mineralogy to distinguish and classify several minerals which occur as single crystals with beautifully developed crystal faces. The method, the tools for measurements were all well developed by mineralogists. It found many applications in metallurgical and materials engineering in later times. Phase transformation studies, nucleation and growth

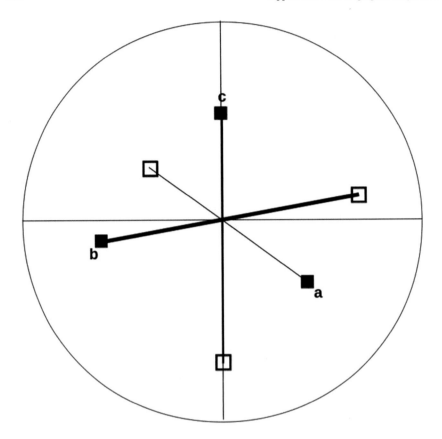

Fig. A.6 Stereographic projection of a typical triclinic crystal

in solid state, orientation of surface marks of twins, slip planes, deformation bands, identification of all possible variants of a precipitate phase in a matrix and more recently in identification of crystallographic orientation (Pole figures) in EBSD are a few examples to cite among numerous applications.

1. Dana, E.S., Ford, W.E.: A Text Book of Mineralogy with an Extended Treatise on Crystallography and Physical Mineralogy, First Indian Edition, pp. 48–63. Asia Publishing House (1959).
2. Barrett, C., Massalski, T.B.: Structure of Metals, 3rd edn., p. 32. McGraw-Hill, Inc. (1966).
3. Penfield, S.L.: Am. J. Sci. **9**(1–24), 115–144 (1901).
4. Johari, O., Thomas, G.: The stereographic Projection and Its Applications, Interscience Publishers, A Division of Wiley (1969).

Appendix B
Basic Crystallography

Crystals have periodic arrangements of atoms or clusters of atoms on an imaginary lattice which is a three-dimensional grid. Such a lattice is called space lattice and it has to meet certain geometric requirements. By definition, every point on the space lattice has to be identical to every other point. Therefore, any point on the lattice can be reached from any arbitrarily chosen origin by a vector $\mathbf{r} = a\hat{x} + b\hat{y} + c\hat{z}$, where $\hat{x}, \hat{y}, \hat{z}$ are unit vectors (see Fig. B.1). Since the crystal is periodic the lattice has to preserve the periodicity and the vector components a, b and c have to be integers. (This is not the case in quasicrystals and thence in the quasiperiodic lattices.) We will further realise that there is an additional condition of preserving the complete set of symmetry elements of the particular crystal structure.

B.1 Unit Cell and Primitive Cell

From the choice of basis vectors **a, b, c** and their mutual angles, it appears that many possibilities exist for the atoms or groups of atoms of the crystal to decorate the lattice points. Seven crystal systems arise by a choice of different basis vectors and their mutual angles while only 14 different ways of decorating these 7 systems are possible. The 14 lattices are called Bravais lattices and are consistent with the symmetries possessed by the 7 crystal systems (Fig. B.1).

B.2 Crystal Systems and Bravais Lattices

When the points of the space lattice are populated with atoms or groups of atoms, several numbers of crystal structures emerge, but the space lattices remain the same 14 Bravais lattice for three-dimensional crystals. Since every point in the lattice has to

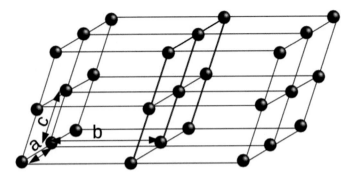

Fig. B.1 A general lattice

Fig. B.2 A typical unit cell

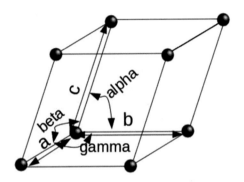

be identical to every other point, a repeating unit has to be identified in the crystal for which again several possibilities exist. Hence, the smallest possible unit is selected which when repeated in three dimensions correctly maps the entire crystal. Such a unit is called a **unit cell** and it possess all the symmetry elements of the crystal. When only one lattice point exists for a unit cell, such a cell is called **Primitive Cell**. Again there is more than one way to choose a primitive cell for a unit cell (Fig. B.2).

As per crystallographic convention, a right-hand rule is followed in defining the axes of the unit cell. According to this convention, the **a–b**-axes constitute the x–y-plane and the angles opposite **a**, **b** and **c** are α, β and γ, respectively.

The unit cell in the figure has one lattice point at each corner while it is also possible to choose a unit cell in such a way that there are more lattice points within the cell. This choice is indeed restricted based on particular symmetry of the crystal. The possibilities are illustrated for each crystal system in the table given below. The number of symmetry elements decreases from cubic crystal to a triclinic crystal. Together with the internal and external symmetry elements of the unit cell, three-dimensional crystals constitute 240 space groups. We are not discussing further about space groups and point groups, except to say that a knowledge of these is essential to understand the intricate details in CBED patterns. Interested readers are referred to standard textbooks devoted to crystallography and electron diffraction listed below (Table B.1).

1. Donald Bloss, F.: Crystallography and Crystal Chemistry. Published by Holt, Rinehart and Winston (1971). ISBN 10:0030851556/ISBN 13: 9780030851551.
2. Vainshtein, B.K.: Structure Analysis by Electron Diffraction. Pergamon Press Ltd., Oxford (1964).
3. Zou, X., Hovmöller, S., Oleynikov, P.: Electron Crystallography-Electron Microscopy and Electron Diffraction. Oxford University Press, New York (2011).

Note: P is Primitive, C is Base Centred (commonly taken as the face perpendicular to C-axis), I is Innocent art, i.e. Body Centred, F is Face Centred and R is Rhombohedral.

Table B.1 Crystal systems and Bravais lattices

Crystal system	Unit cell	Cell parameters	Bravais lattices
Cubic		$a = b = c$, $\alpha = \beta = \gamma = 90°$	P,I,F
Hexagonal		$a = b \neq c$, $\alpha = \beta = 90°$, $\gamma = 120°$	P
Rhombohedral		$a = b = c$, $\alpha = \beta = \gamma \neq 90°$	R
Tetragonal		$a = b \neq c$, $\alpha = \beta = \gamma = 90°$	P, I
Orthorhombic		$a \neq b \neq c$, $\alpha = \beta = \gamma = 90°$	P, C, (A, B), F, I
Monoclinic		$a \neq b \neq c$, $\alpha = \gamma = 90°$, $\beta \neq 90°$ or $\alpha = \beta = 90°$, $\gamma \neq 90°$	P, C, (A, B)
Triclinic		$a \neq b \neq c$, $\alpha \neq \beta \neq \gamma \neq 90°$	P

Appendix C
Cosine Formulae

C.1 Table of Cosines

Table C.1 Cosine formulae

Crystal system, Angle ϕ between (h_1, k_1, l_1) and (h_2, k_2, l_2) Bravais lattices

Cubic

$$\cos\phi = \frac{h_1 h_2 + k_1 k_2 + l_1 l_2}{\sqrt{[(h_1^2 + k_1^2 + l_1^2)(h_2^2 + k_2^2 + l_2^2)]}}$$

Tetragonal

$$\cos\phi = \frac{\dfrac{1}{a^2}(h_1 h_2 + k_1 k_2) + \dfrac{1}{c^2}(l_1 l_2)}{\sqrt{[[\dfrac{1}{a^2}(h_1^2 + k_1^2) + \dfrac{1}{c^2}(l_1^2)][\dfrac{1}{a^2}(h_2^2 + k_2^2) + \dfrac{1}{c^2}(l_2^2)]}}$$

Orthorhombic

$$\cos\phi = \frac{\dfrac{1}{a^2}h_1 h_2 + \dfrac{1}{b^2}k_1 k_2 + \dfrac{1}{c^2}l_1 l_2}{\sqrt{[[\dfrac{1}{a^2}h_1^2 + \dfrac{1}{b^2}k_1^2 + \dfrac{1}{c^2}l_1^2][\dfrac{1}{a^2}h_2^2 + \dfrac{1}{b^2}k_2^2 + \dfrac{1}{c^2}l_2^2]}}$$

Hexagonal

$$\cos\phi = \frac{h_1 h_2 + k_1 k_2 + \frac{1}{2}(h_1 k_2 + k_1 h_2) + \frac{3}{4}\dfrac{a^2}{c^2}l_1 l_2}{\sqrt{[(h_1^2 + k_1^2 + h_1 k_1 + \frac{3}{4}\dfrac{a^2}{c^2}l_1^2)(h_2^2 + k_2^2 + h_2 k_2 + \frac{3}{4}\dfrac{a^2}{c^2}l_2^2)]}}$$

(continued)

G. V. S. Sastry, *Microstructural Characterisation Techniques*,
Indian Institute of Metals Series, https://doi.org/10.1007/978-981-19-3509-1

Table C.1 (continued)

Rhombohedral cell can be re-indexed on Hexagonal indexing scheme and use the same expression as in Hexagonal

Monoclinic

$\text{Cos}\phi =$

$$\frac{\frac{1}{a^2}h_1h_2 + \frac{1}{b^2}k_1k_2 Sin^2\beta + \frac{1}{c^2}l_1l_2 - \frac{1}{ac}(l_1h_2 + l_2h_1)Cos\beta}{\sqrt{[[\frac{1}{a^2}h_1^2 + \frac{1}{b^2}k_1^2 Sin^2\beta + \frac{1}{c^2}l_1^2 - \frac{2h_1l_1}{ac}Cos\beta][\frac{1}{a^2}h_2^2 + \frac{1}{b^2}k_2^2 Sin^2\beta + \frac{1}{c^2}l_2^2 - \frac{2h_2l_2}{ac}Cos\beta]]}}$$

Triclinic

$\text{Cos}\phi = \frac{nume}{denom1 x denom2}$

$nume = h_1h_2b^2c^2 Sin^2\alpha + k_1k_2a^2c^2 Sin^2\beta + l_1l_2a^2b^2 Sin^2\gamma + abc^2(Cos\alpha Cos\beta - Cos\gamma)(k_1h_2 + h_1k_2)$

$+ab^2c(Cos\gamma Cos\alpha - Cos\beta)(h_1l_2 + l_1h_2)$

$+a^2bc(Cos\beta Cos\gamma - Cos\alpha)(k_1l_2 + l_1k_2)$

$denome\ 1 = \sqrt{\Big[h_1^2b^2c^2 Sin^2\alpha + k_1^2a^2c^2 Sin^2\beta + l_1^2a^2b^2 Sin^2\gamma }$

$+ 2h_1k_1abc^2(Cos\alpha Cos\beta - Cos\gamma)$

$+ 2h_1l_1ab^2c(Cos\gamma Cos\alpha - Cos\beta)$

$+ 2k_1l_1a^2bc(Cos\beta Cos\gamma - Cos\alpha)\Big]$

$denome\ 2 = \sqrt{\Big[h_2^2b^2c^2 Sin^2\alpha + k_2^2a^2c^2 Sin^2\beta + l_2^2a^2b^2 Sin^2\gamma }$

$+ 2h_2k_2abc^2(Cos\alpha Cos\beta - Cos\gamma)$

$+ 2h_2l_2ab^2c(Cos\gamma Cos\alpha - Cos\beta)$

$+ 2k_2l_2a^2bc(Cos\beta Cos\gamma - Cos\alpha)\Big]$

Appendix D
Illumination Systems for Optical Microscopes

Source of illumination for a bench microscope used for examination of biological or mineralogical transparent samples can be simple sunlight reflected by a concave mirror. In the case of opaque metallurgical or material samples, we need special arrangement for the light source which can be either the simple tungsten lamp or the most advanced LED lighting. In either case of samples, we need Köhler illumination for a uniformly illuminated image. Importance of Köhler illumination, particularly in the case of a metallograph, has been explained in Chap. 6. Under this type of illumination, the light source is so placed before the condenser lens that the field aperture and consequently the field of view are uniformly illuminated and are kept independent of condenser lens. This arrangement in turn demands uniformity of emission of light from the lamp used. Not all available light sources (lamps) are able to fulfil this requirement of Köhler illumination.

The arc lamps, not only are of high brightness, are suitable for fluorescence microscopy. When compared with them, the tungsten-halogen filament lamps are less radiant because of the spread of the source (i.e. filament). Arc works as an ideal point source. The latest models of microscopes employ LED illumination for the same reason of functioning as ideal point source of monochromatic light. Their brightness is indeed much less compared to the spectral peaks of a mercury HBO 100-watt arc lamp. High-power laser light sources, used specifically in confocal laser microscopy, excel in their radiant energy in comparison to arc lamps, tungsten-halogen incandescent lamps or LEDs.

D.1   Important Parameters

Colour temperature is one of the important parameters that we need to understand. It is the temperature expressed in Kelvin, of a perfect black body which emits radiation similar to the radiation of our light source under consideration. Thus, it is specifically

G. V. S. Sastry, *Microstructural Characterisation Techniques*,
Indian Institute of Metals Series, https://doi.org/10.1007/978-981-19-3509-1

relevant to the tungsten lamps and tungsten-halogen lamps which are of thermal radiation type. The range of wavelengths over which the filament is emitting radiation is an important parameter in colour microscopy and recording with colour film. The film is more sensitive to certain colours and its sensitivity is also expressed in terms of colour temperature. So is the response from photographic paper. For best representation of the colour contrast of the specimen, the colour temperatures of the light source, photographic film and the print paper should be the same or nearly matching. It is somewhat counter-intuitive to know that the yellowish colour of the incandescent tungsten lamp is at a lower colour temperature (more towards infrared) compared to that of blue light which is at a higher colour temperature.

Currently, we use computer monitors to display the images and digital cameras to acquire those. In addition, the light sources used in the modern microscopes are of non-radiative type like LEDs, lasers and arc lamps. Therefore, a correlated colour temperature, CCT, is defined to take into account the non-radiative type of devices, i.e. whose radiations are not related to black-body emissions directly, which are thermal in nature, but their chromaticity index is close to that of the black-body radiation.

D.1.1   Mercury Arc Lamps

Mercury arc lamps are extensively used in fluorescence microscopy owing to the close proximity of its spectral lines with the excitation maxima of the fluorophores. Xenon arc lamps are another variety with uniform continuous spectral lines of the same class. Arc lamps do take time to reach steady-state emission after switching on and generate excessive heat.

D.1.2   Metal Halide Lamps

These are as well arc discharge type in which the metal halide lamp, which is usually metal iodide salts, is housed in an elliptical reflector and the light is delivered at the microscope through a light pipe (a liquid light guide not the fibre-optic type). They are extremely bright and compete well with mercury or xenon arc lamps.

D.1.3   Light Emitting Diodes

The LEDs form the most promising development of illumination source for optical microscopy. A high-power diode generates a coherent monochromatic light beam. The wavelength of the emitted light can be manipulated by a suitable choice of the LED. These lamps are very useful in fluorescence microscopy.

References

Andrews, K.W., Dyson, D.J., Keown, S.R.: Interpretation of Electron Diffraction Patterns. Adam Hilger Ltd., London (1971)

Ash, E.A., Nicholls, G.: Nature **237**, 510–512 (1972)

Ashby, M.F., Brown, L.M.: Phil. Mag. **8**, 1083(a) and 1649(b) (1963)

Ayache, J., Beaunier, L., Boumendil, J., Ehret, G., Laub, D.: Sample Preparation Handbook for Transmission Electron Microscopy: Methodology and Techniques, vols. 1 and 2. Springer (2010, 2014)

Betzig, E., Trautman, J.K., Harris, T.D., Weiner, J.S., Kostelak, R.L.: Science **251**(5000), 1468–1470 (1991)

Buxton, B.F., Eades, J.A., Steeds, J.W., Rackham, G.M.: Philos. Trans. R. Soc., London, Ser. A. **281**, 171 (1976)

Cahn, J.W., Shechtman, D., Gratias, D.: J. Mater. Res. **1**, 13–26 (1986)

Chen, J., Hirayama, T., Lai, G., Tanji, T., Ishizuka, K., Tonomura, A.: Opt. Lett. **18**, 1887–1889 (1993)

Cherns, D., Preston, A.R.: Convergent beam diffraction studies of crystal defects In: Imura et al. T., (eds.), Proceedings of the 11th International Congress on Electron Microscopy, Kyoto, Japan, vol. 1, pp. 207–208. The Japanese Society of Electron Microscopy, Tokyo, Japan (1986)

Cherns, D., Morniroli, J.-P.: Ultramicroscopy **53**, 167–180 (1994)

Cowley, J.M.: In: High Resolution Transmission Electron Microscopy and Associated Techniques, Buseck, P.R., Cowley, J.M., Eyring, L. (eds.), pp. 1–34. Oxford University Press (1992)

Cowley, J.M.: Acta Crystall. **24**, 557–563 (1967)

de Graef, M.: Introduction to Conventional Transmission Electron Microscopy. Cambridge University Press (2003)

Fultz, B., Howe, J.: Transmission Electron Microscopy and Diffractometry of Materials, 4th edn. Springer (2013)

Goodman, P.: Acta Crystallogr. A **31**, 79 (1975)

Groves, T.R.: Charged Particle Optics Theory: An Introduction. CRC Press, Taylor and Francis (2014)

Hawkes, P.W., Kasper, E.: Wave Optics, vol. 3. Academic Ltd. London, Academic Inc., San Diegao (1994)

Hawkes, P.W., Kasper, E.: Principles of Electron Optics, Vol 1, Basic Geometrical Optics, Academic Press, 2nd Ed., (2017)

Hirth, J.P., Lothe, J.: Theory of Dislocations. Wiley (1982)

Jagannathan, R., Simon, R., Sudershan, E.C.G., Mukunda, N.: Phys. Lett. A **134**, 457 (1989)

Jagannathan, R., Simon, R., Sudershan, E.C.G., Mukunda, N.: Phys. Rev. A **42**, 6674 (1990)

© The Editor(s) (if applicable) and The Author(s), under exclusive license to Springer Nature Singapore Pte Ltd. 2022
G. V. S. Sastry, *Microstructural Characterisation Techniques*, Indian Institute of Metals Series, https://doi.org/10.1007/978-981-19-3509-1

Jaiswal, R., Agarwal, K., Pratap, V., Soni, A., Kumar, S., Mukhopadhyay, K., Eswara Prasad, N.: Mater. Sci. Eng. B **262**, 114711 (2020)

Johari, O., Thomas, G.: The Stereographic Projection and Its Applications. Interscience Publishers. A Division of Wiley & Inc., New York (1969)

Kainuma, Y.: Acta Cryst. **8**, 247–257 (1955)

Leith, E.N., Upatneiks, J.: J. Opt. Soc. Am. **52**, 1123–1130 (1962)

Lipson, S.G., Lipson, H.: Optical Physics, 1st edn. Cambridge University Press (1969)

Loretto, M.H.: Electron Beam Analysis of Materials. Chapman and Hall, London (1984)

Lu, P.J., Deffeyes, K., Steinhardt, P.J., Yo, N.: Phys. Rev. Lett. **87**, 275507 (2002)

Mandal, R.K., Pramanick, A.K., Sastry, G.V.S., Lele, S.: Materials Research Society Symposium Proceedings, vol. 805. Materials Research Society (2004)

Mandal, R.K., Lele, S.: Phys. Rev. Lett. **62**, 2695 (1989)

Manna, R.: Ph.D. Thesis: Design and Characterisation of Ultrafine Grained Aluminum and Aluminum Alloys Processed by Equal Channel Angular Pressing. IIT (BHU), Varanasi (2008)

Mordina, B., Tiwari, R.K., Setua, D.K., Sharma, A.: J. Phys. Chem. C **118**, 25684–25703 (2014)

Morin, D.: In: Waves, Chapter 3 (Version 1, November 28, 2009). https://scholar.harvard.edu/files/david-morin/files/waves-fourier.pdf

Morniroli, J.P., Stadlemann, P., Ji, G.: Nicolopoulos. J. Microsc. **237**, 511–515 (2010)

Pluta, M.: Advanced Light Microscopy: Specialised Methods, vol. 2. Elsevier, Amsterdam (1988)

Ramachanrarao, P., Sastry, G.V.S.: Pramana. J. Phys. **25**, L225–L230 (1985)

Reimer, L., Kohl, H.: Transmission Electron Microscopy: Physics of Image Formation, 2nd edn. Springer Science+Business Media, LLC, Berlin (2008)

Rose, H.H.: Geometrical Charged-Particle Optics, 2nd edn. Springer, Berlin (2009)

Russ, J.C., Neal, F.B.: The Image Processing Handbook, 7th edn. CRC Press (2016)

Sarath kumar, G.V., Mangipudi, K.R., Sastry, G.V.S., Singh, L.K., Dhanasekaran, S., Sivaprasad, K.: Nat. Sci. Rep. **10**, 354–363 (2020)

Sastry, G.V.S., Sudhir, S., Gupta, A., Murty, B.S.: Milling of oxide blends–a possible clue to the mechanism of mechanical alloying. In: Singh, A.K., (ed.), Proceedings of the "Seminar on Advanced X-ray Techniques in Research and Industry(XTRI-2003)". Capital Publisher Company, New Delhi (2005)

Sastry, G.V.S., Suryanarayana, C., Van Sande, M., Van Tendeloo, G.: Mat. Res. Bull. **13**, 1065–1070 (1978)

Sharma, K., Garai, D., Gupta, A., Roy, D., Eswara Prasad, N.: J. Phys. Chem. C **125**, 21653–21662 (2021)

Shechtman, D., Blech, I., Gratias, D., Cahn, J.W.: Phys. Rev. Lett. **53**, 1951 (1984)

Spence, J.C.H., Zuo, J.M.: Electron Microdiffraction. Springer Science+Business Media, LLC (1992)

Spence, J.C.H.: High-Resolution Electron Microscopy, IV edn. Oxford University Press (2013)

Tanaka, M., Kaneyama, T.: CBED studies of imperfect crystals In: Imura et al. T., (eds.), Proceedings of the 11th International Congress on Electron Microscopy, Kyoto, Japan, vol. 1, pp. 203–204. The Japanese Society of Electron Microscopy, Tokyo (1986)

Tanaka, M., Terauchi, M., Kaneyama, T.: Convergent Beam Electron Diffraction II. JEOL, Tokyo (1988)

Tonomura, A.: Electron Holography. Second Enlarged Edition. Springer (1999)

Vincent, R., Midgley, P.A.: Ultramicroscopy **53**, 271–82 (1992)

Vincent, R., Midgley, P.A.: Ultramicroscopy **53**, 271–282 (1994)

Weisenburger, S., Sandoghdar, V.: Contemp. Phys. **56**(2), 123–143 (2015)

Williams, D.B., Barry Carter, C.: Transmission Electron Microscopy A Textbook for Materials Science. Springer Science+Business Media, LLC (2009)

Zhang, D.L., Grüner, D., Oleynikov, P., Wan, W., Hovmöler, S., Zou, X.D.: Ultramicroscopy **111**, 47–55 (2010)

Zou, X., Hovmöller, S., Oleynikov, P.: Electron Crystallography: Electron Microscopy and Electron Diffraction. IUCR Texts on Crystallography-16. Oxford Science Publications (2011)

Printed in the United States
by Baker & Taylor Publisher Services